Praise for
Microbe Science for Gardeners

Science guides me in everything I do with gardening.
It all starts in the soil, but not without microbes! Yet, understanding the complexities of the many ways they make soil better
and plants healthier can seem overwhelming. Robert Pavlis'
newest book, *Microbe Science for Gardeners* beautifully breaks it
down in his usual no-nonsense way. If you want to learn anything
about the science of soil, and what's in it, this book will help
you easily understand the vital role microbes play in bringing
soil (and everything growing in it) to life!

—Joe Lamp'l, founder, joegardener.com, The Online Gardening
Academy™, creator/ executive producer, Growing a Greener World®

Robert Pavlis' *Microbe Science for Gardeners* is an accessible
and understandable dive into the amazing relationship between
microbes and plants. Useful and practical gardening advice.

—Jeff Lowenfels, author, the *Teaming Series*
and *DIY Autoflowering Cannabis*

It's hard not to appreciate Pavlis' clear and consistently
myth-busting approach to the subject, guiding the reader with
a precise balance of science and analogy. There is loads of good
information coupled with actionable insight throughout.
It's deeply in-tune with the current science and Pavlis also
introduces us to lesser-known and developing ideas in soil
microbiology that we will see grow over the next few years.
No matter your understanding of soil biology, I suspect
Microbe Science for Gardeners will leave you deeply enriched,
as it has me. From backyard to farm-scale, this is a book
every grower should possess.

—Jesse Frost, author, *The Living Soil Handbook: The No-till Grower's
Guide to Ecological Market Gardening*

This fascinating book by Robert Pavlis presents the latest science on what is known (and not known) about the multitude of bacteria, fungi, and other microbes that fill the soil, cover leaves, and interact with roots. Read it for the wealth of well-researched, practical information that will make you a successful and savvy gardener—but also read it to be delighted by the wonders of this astonishing invisible world. This is an outstanding resource and should be in the hands of anyone interested in gardening!

—Linda Gilkeson, author, *Backyard Bounty: The Complete Guide to Year-Round Organic Gardening in the Pacific Northwest*

A must-have for gardeners who want to understand the unseen worlds above and below the soil, Robert Pavlis' latest book explains the hidden life keeping our plants thriving.

—Rebecca Martin, technical editor, *Mother Earth News*

Robert Pavlis has a unique ability to communicate complex topics with ease and clarity. He has done it again with *Microbe Science for Gardeners*. This comprehensive exploration of soil ecology examines the dynamic relationships between plants and the microbiome of the soil, providing the reader with a guide to understanding the role of bacteria, viruses, fungi and other microbes in the creation and maintenance of healthy soil.

—Darrell Frey, owner, Three Sisters Farm, and author, *Bioshelter Market Garden: A Permaculture Farm*

microbe science
for gardeners

Secrets to Better Plant Health

ROBERT PAVLIS

new society PUBLISHERS

Inquiries regarding requests to reprint all or part of *Microbe Science for Gardeners* should be addressed to New Society Publishers at the address below. To order directly from the publishers, please call 250-247-9737 or order online at www.newsociety.com.

Any other inquiries can be directed by mail to:

New Society Publishers
P.O. Box 189, Gabriola Island, BC V0R 1X0, Canada
(250) 247-9737

LIBRARY AND ARCHIVES CANADA CATALOGUING IN PUBLICATION
Title: Microbe science for gardeners : secrets to better plant health /
Robert Pavlis.
Names: Pavlis, Robert, author.
Description: Includes bibliographical references and index.
Identifiers: Canadiana (print) 20230459412 | Canadiana (ebook)
20230459617 | ISBN 9780865719774 (softcover) | ISBN 9781550927719
(PDF) | ISBN 9781771423670 (EPUB)
Subjects: LCSH: Plant-microbe relationships. | LCSH: Microorganisms. |
LCSH: Gardening.
Classification: LCC QR351 .P38 2023 | DDC 579/.178—dc23

Funded by the Government of Canada | Financé par le gouvernement du Canada | **Canadä**

New Society Publishers' mission is to publish books that contribute in fundamental ways to building an ecologically sustainable and just society, and to do so with the least possible impact on the environment, in a manner that models this vision.

FSC MIX Paper from responsible sources FSC® C016245

Certified B Corporation

new society PUBLISHERS

Contents

We know more about the movement of celestial bodies than about the soil underfoot.

—Leonardo da Vinci

The microbes in one acre of soil weigh the same as two cows.

—unknown

Great fleas have little fleas upon their backs to bite 'em,
And little fleas have lesser fleas, and so ad infinitum.

—Augustus de Morgan

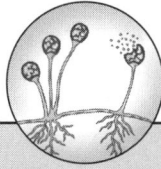

CHAPTER 1

Introduction

WHAT DO YOU SEE when you look at the surface of a leaf?

The surface is mostly smooth, but it can have some bumps on it and some leaves are quite hairy. The color is mostly green, although yellow and red also concur on garden plants. You might even see an insect or two crawling across the leaf, but other than that, there is not much activity.

Your perspective of that leaf is very wrong because you are using macro eyes. They just don't see the details very well. If you look at the leaf with a microscope, you suddenly see a whole new world that is full of millions of organisms. Some are stationary and others are speeding along. Admittedly, speeding along at microscopic levels is actually quite slow.

Not only do you see many individual organisms, but you also see many different types of organisms. Some, like viruses, are extremely small and can't be seen even with a light microscope. In contrast, others are relatively huge, multi-celled organisms. Even shapes vary a lot. You'll see spheres, long rods, and undefined blobs. Some are whipping hair-like appendages around to help them motor along.

The colors are fabulous. Some are clear with almost no coloration, but many have shades of blue, green, red, and violet. Their surfaces also vary a lot, and scientists use this texture to help with identification.

It might seem like an idyllic environment. All natural and cozy, but it is anything but. Microbes are constantly fighting for food and space. Small ones are eaten by larger ones, who get eaten by even larger ones. These guys don't even fight fairly and use all kinds of chemical weapons to destroy each other.

This is one very complex society, and to be honest, scientists are just starting to understand it.

The most common microbes are bacteria, and a gram of fresh leaf, the weight of a paperclip, may harbor as many as a hundred million of them.

Let's dig a little deeper. Scrape off all of the surface microbes so that we can see the leaf. Under a microscope, the surface no longer looks smooth. It is mountainous with all kinds of valleys, cracks, and holes. These are perfect places for microbes to hide.

If you look a little closer at the holes, you will find some very large ones, the stomata. The plant uses these to absorb carbon

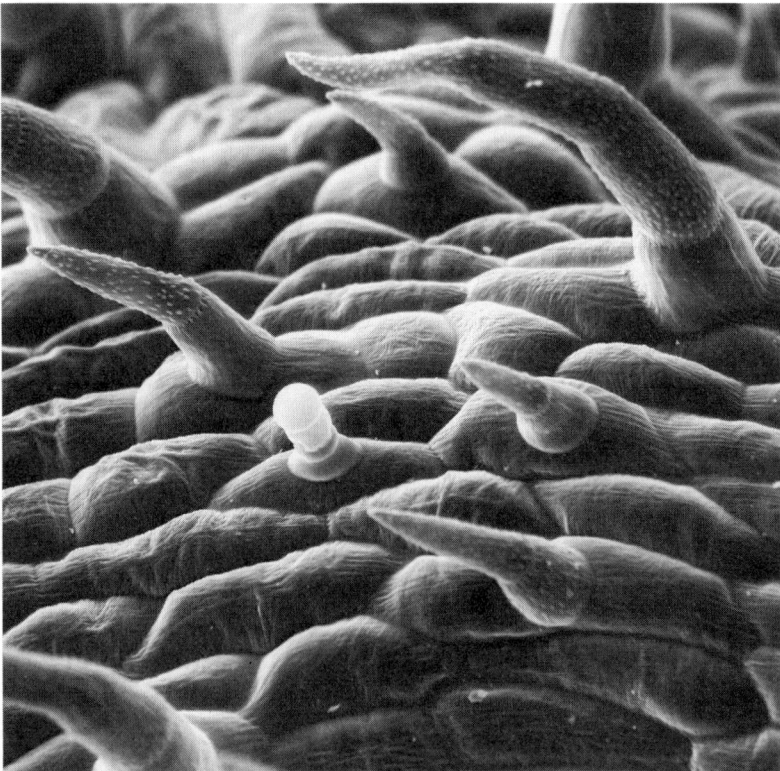

Surface of a coleus leaf

dioxide and expel excess oxygen, water, and other gases. The microbes take full advantage of these and crawl right inside the leaf. Some spend their whole life inside leaves.

Once inside the plant, microbes can even enter plant cells. Some of these are very beneficial to plants, who actually send out chemical signals to attract them. Others, like viruses, can be quite harmful.

A leaf is covered in thousands of different microbes. Some are beneficial, some are neutral, and others are harmful pathogens. This book is all about these microbes and their interaction with plants and each other.

Why Learn About Microbes?

Why learn about microbes when you could be learning how to care for plants? Would that not make you a better gardener? Perhaps, but one thing I have learned after gardening for many years is that learning about plants only takes you so far. There are just too many plants to study. I have learned that if you take the time to understand the underlying basis of nature, growing any plant becomes easy.

Microbes are vital to plant growth. They help plants get nutrients from soil and dead organic matter. They cover every square inch of the plant, including leaves, stems, flowers, fruits, and even the roots. Some are beneficial to plants, others are pathogens ready to kill the plant, and many play a neutral role. But even these neutral actors are critical for soil structure, soil nutrient levels, and plant health.

Understanding the interaction between plants and microbes is as important as learning how to water your plants or how to situate them correctly for the right amount of light. You can't see the microbes, but they are everywhere, and everything you do in the garden affects them and in turn your plants.

As we travel down the road of understanding, you will learn about microbes that plants farm to get more nitrogen. Plants also allow microbes to "pollinate" flowers so that they end up in seeds to help future generations fight bacterial infections. Special fungi attach themselves to roots to extend the plant's reach in soil, making it easier to find nutrients.

Plant-available phosphate is a rare resource in soil, and microbes collect it for plants. Nitrogen-fixing bacteria take nitrogen gas from the air and convert it to a form plants can use. But did you know that it is the plant that initiates and manages these associations? Plants actively manipulate the microbe community around themselves.

Gardeners become obsessed with plant diseases, and microbial pathogens are certainly important. What is more surprising to me is that most diseases are preventable, not by direct actions of the gardener but by the activity of invisible microbes in and on the plants.

How many species of living things inhabit earth? That seems like a simple question, but we still don't know the answer because most species have not yet been identified. We have named around 1.5 million of them. About two thousand new native plants are discovered every year. There are many spots on earth that have never been botanized, so the number is certain to grow.

The largest gap in our understanding of organisms is with microbes. Their small size and visible similarities make it very difficult to identify species. It is only now with the help of DNA analysis that we are starting to appreciate their numbers.

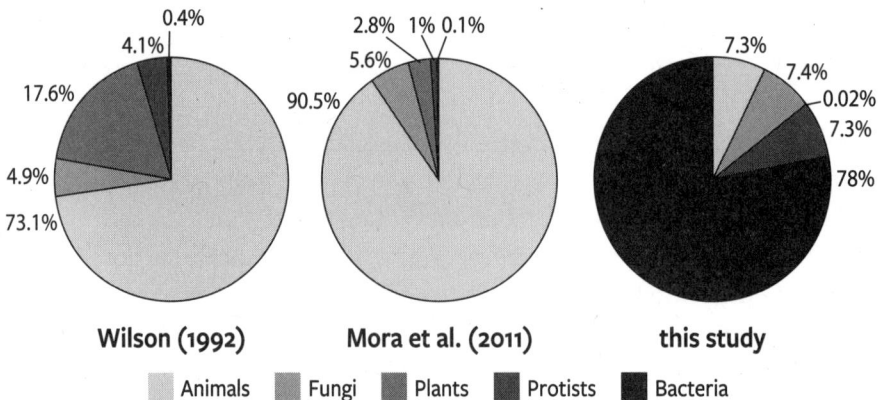

Pie of Life: Relative number of species on earth. Credit: John J. Wiens et al.[1]

Armed with new DNA data, scientists have developed a new estimate of life on earth that is between one and six billion species. These results are still quite speculative, but they are changing our understanding of our world. The "Pie of Life" charts show how our estimates have changed over time. The latest estimate shows that microbe species dominate (70–90 percent) of the planet.

Terms Used in This Book

Science is full of weird terms, but it is critical to know these terms in order to understand the underlying meaning. I have kept technical terms to a minimum, but a few are important for gardeners to know.

Plant Spheres

Scientists have defined specific microbe ecosystems as spheres. I have used the terms phyllosphere and rhizosphere in this book:

- anthosphere—area around a flower
- carposphere—area around a fruit
- phyllosphere—area around leaves and stems
- rhizosphere—area around roots

Epiphytes and Endophytes

An epiphyte is an organism that grows on the surface of a plant. Orchids are a good example of this. The term is also used to describe microbes that grow on plants.

An endophyte is an organism that lives inside a plant. Endophytic microbes are found in all parts of a plant, including the seeds.

Strain VS Species

The terms genus and species are used to identify a particular organism and they work very well for larger organisms such as animals and plants. Microbes add an extra complication because they mutate very quickly and they can easily exchange pieces of DNA. For any given species, there can be numerous variations and rather than identifying each one as a new species, scientists tend to call them strains. Strains are different enough to be identifiable but not different enough to warrant a species designation.

Facultative Anaerobe

All organisms require energy to live and those living in an oxygen-rich environment, like most animals and plants, use oxygen to make an energy molecule called ATP—the energy battery of living things. Such organisms are called aerobic.

Some organisms, mostly microbes, live in an oxygen-poor environment and get their energy through fermentation. These are called anaerobic.

There is a third class of organisms called facultative anaerobes. These guys will get their energy using ATP when oxygen is plentiful but can switch to fermentation when the oxygen level drops. Yeast is a facultative anaerobe, as are bacteria such as *Staphylococcus* spp., *Escherichia coli*, and *Salmonella*.

This ability to live in both environments makes it easier for them to survive as conditions change. You might recognize *E. coli* as a common gut bacteria that grows in our intestine, which is a low-oxygen environment, but it also lives in soil and on leaves, which are high-oxygen environments.

Many gardeners make the mistake of thinking that pathogens only exist in anaerobic conditions, but all three of the above-mentioned bacteria can cause disease and infection.

Eukaryotes and Prokaryotes

The cells of all living organisms can be divided into one of two categories: eukaryotic and prokaryotic.

Eukaryotes are organisms that have eukaryotic cells, which are the basis of all multicellular organisms including animals, plants, and fungi. These cells have well-defined organelles inside their cells. Examples include the nucleus that contains the DNA or RNA, the endoplasmic reticulum, which is used to synthesize proteins, lipids, and steroids, and the mitochondria, which produces energy.

How do you remember which is which? Simple. You are a complex organism and therefore a "Youkaryote."

Beneficial VS Effective Microbes

These two terms are used a lot in gardening discussions and are often misused. They are not the same thing.

Beneficial Microbes

Most academic discussions are just about microbes, but gardeners and manufacturers of products like to use the term "beneficial microbes." By definition, these are microbes that are beneficial to plants and include anything that is not a pathogen. Even if a microbe does not help a plant directly, it usually helps indirectly by providing nutrients to soil, or competing with pathogens.

The term *beneficial* does not really tell you much about the microbe, except that it is not a pathogen. Ignore this term on product labels—it is just marketing gibberish.

Effective Microbes

The term *Effective Microorganisms* (EM) was first used by Dr. Teruo Higa to describe a combination of about eighty different microbes that were capable of improving the decomposition of organic matter. He developed the idea that the right combination of "positive microbes" would improve any media, including soil. The initial product was called EM-1, which contained three groups of microbes: yeast, photosynthetic bacteria, and lactic acid bacteria.

Since the introduction of EM-1, many other formulations have been produced by a variety of manufacturers. When you buy a product containing EM, you are buying a combination of microbes that the manufacturer considers important.

Fertilizer

Most gardeners use the term *fertilizer* to refer to synthetic fertilizer, but once you understand that the nutrients from both synthetic fertilizer and organic fertilizer are identical, you realize that both will have the same effect on plants and microbes. It is the amount of added nutrients that is key, not the type.

In this book, I use the term fertilizer to refer to both synthetic and organic fertilizer. If it is important to differentiate between the two, I'll call them synthetic fertilizer and organic fertilizer, with the latter referring to a wide range of products including manure, compost, blood meal, etc.

Miscellaneous Terms

Bulk Soil—the soil outside of the rhizosphere, which includes most of the garden soil.

Microbiome—the microorganisms in a particular environment, such as the surface of a leaf (the leaf microbiome) or the inside of your gut (the gut microbiome).

Mineralization—the conversion of organic matter to inorganic minerals (i.e., the creation of minerals/nutrients).

Immobilization—the conversion of inorganic minerals to organic matter (i.e., minerals become incorporated into organic molecules).

Abiotic—nonliving factors such as moisture level, temperature, and soil type.

Symbiont—an organism that is very closely associated with another, usually larger, organism. This larger organism is called a host.

Micrometer (μm)—a useful unit for measuring size in the microbe world. One inch equals 25,400 μm. One cm = 1,000 mm = 10,000 μm.

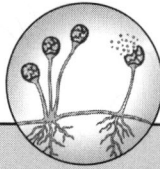

CHAPTER 2

The World Under a Microscope

Microbes by the Numbers

YOU CAN'T SEE THEM or touch them but microbes exist in vast numbers. It has been suggested that the number of bacteria on earth is 5,000,000,000,000,000,000,000,000,000,000. This is five million trillion trillion, or 5×10^{30}.

To make it easier to understand this large number, consider just one gram of soil. That is the weight of a single paper clip, or the amount of soil under your fingernails after an hour of gardening.

The table of microbes shows some of the most common microbe categories and their respective numbers in one gram of garden soil.

	Size (µm)	Rate of Reproduction	Number/g of soil	biomass (g/m²)
Bacteria	1–10	20 minutes	10^9	40–500
Actinomycete	1–2	varies	10^8	40–500
Fungi	2–10 wide 5–50 long	varies	10^7	100–1,500
Algae	1–2	24 hours–4 weeks	10^6	1–50
Protozoa	10–100	4–8 hours	10^5	> 2,000
Nematode	300–5,000	30 days	10^4	varies
Virus	0.03			
Human hair	40–70			

Microbe Number and Biomass in the top 6 inches of soil, adapted from Hoorman 2010.[2]

It contains a billion bacteria and a million fungi. There can even be more than a thousand nematode worms.

The right-hand column is the biomass (total weight of the organism). You will note that although there are fewer fungi in soil, their mass can actually be larger than that of bacteria. In general, mass equates to biological activity. Organisms with a larger mass eat more, use up more oxygen, produce more CO_2, and have higher amounts of waste. Mass affects the ecosystem more than the number of individual organisms.

These numbers are so large they are hard for us to understand, so this might help: the microbes in one acre of soil weigh the same as two cows.

Microbes Are Important to Plants

Plants can be grown in hydroponic conditions where they are supplied with fertilizer, water, and light. This is a very protected environment where pests and diseases are restricted and closely controlled, but even here plants have microbes on them.

In nature, microbes replace the protected environment of a greenhouse. They cycle nutrients and feed the plants. They improve soil conditions so that water is more readily available, and they protect plants from disease.

You are not just growing plants. You are growing a whole microbial community that in turn helps you take care of the plants.

Plant Growth-Promoting Microorganisms

Plant growth-promoting microbes (PGPM) are organisms that are beneficial to plants. This is a very general definition and as such includes many of the microbes discussed in this book. The benefits include:

- nitrogen fixation
- production of plant hormones such as auxins, gibberellins, and cytokinins
- synthesis of vitamins and antibiotics

- mineral solubilization
- degradation of toxins

An interesting example of PGPM are the 538 yeast strains found under chestnut trees. Seventy-seven of them synthesize indole-3-acetic acid (IAA), with fifteen producing high levels. IAA is the main hormone (an auxin) in plants and it regulates several important processes including growth, cell division, tissue differentiation, apical dominance, and responses to light, gravity, and pathogens. Roots are very sensitive to IAA levels. In addition to helping the trees grow better, the yeast also provides antifungal protection.

Many gardeners have used IBA, a synthetic form of IAA, as a rooting hormone on cuttings.

Energy Food Web

All living organisms require a source of energy. The ones that can use the sun as an energy source are called autotrophs, and those that use carbon compounds are called heterotrophs. A few bacteria that get their energy from minerals are also called autotrophs.

Plants, algae, and a few specialized bacteria, such as cyanobacteria, are autotrophs that can produce their own carbon food using photosynthesis. They use the energy of the sun to convert CO_2 from the air into energy-rich compounds containing carbon.

Heterotrophs can't make their own food. They get their energy needs met by eating organisms that contain compounds such as sugars, fats, and carbohydrates. We get our energy from these carbon compounds.

The Energy Food Web diagram shows how carbon moves from plants to most other life forms. Carbon is passed from one to the other as each life form consumes its food. When an organism dies, the organic matter is used by microbes as their energy source and they in turn become the food source for larger heterotrophs. The carbon moves from organism to organism until it again becomes dead organic matter and the cycle repeats all over again. The key point here is that all of the carbon energy originates with the autotrophs: the plants and algae.

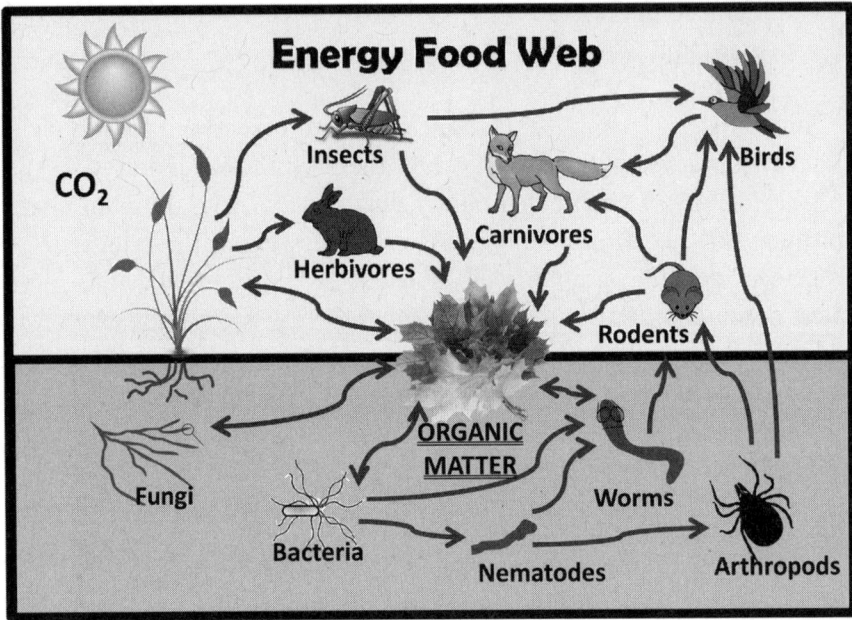

Energy Food Web

Sun
CO_2
Insects
Birds
Carnivores
Herbivores
Rodents
Fungi
ORGANIC MATTER
Worms
Bacteria
Nematodes
Arthropods

If the energy food web continued as I have described it, the amount of organic carbon compounds on earth would be continually increasing, but that is not what happens. As organisms extract the energy out of the food, they produce CO_2 through a process called respiration. Animals get rid of excess CO_2 by breathing it out. Plant roots also respire and expel CO_2. A significant amount of CO_2 is also created during composting as microbes digest the dead organic matter.

The Reality of Composting

When we throw an apple core on the ground it decomposes. If we put the same apple core on a pile, we call it composting. I prefer using the term composting because gardeners are more familiar with it but from a chemical perspective, both processes are identical.

Composting starts with the raw materials. These are normally organic materials that you would easily recognize. An apple core, a banana peel, a dead tomato plant, grass clippings, and some newspaper. You can recognize each of these input materials because they are relatively large.

If you look at these materials on a cellular level, you would see a lot of small details. The stem of the tomato plant has xylem and phloem tubes which are surrounded by epidermis (outer skin). The inside of a leaf has an intricate structure of different layers. Everything is made up of cells, which have their own internal organs.

Now let's look at this more closely on a molecular level. Everything is made out of complex molecules. Big proteins containing thousands of atoms. DNA strands that are huge. Complex starches and oils are made from long chains of atoms.

The difference between a live tomato plant and a newly dead one on a molecular level is zero. There has been virtually no decomposition in that dead tomato plant, and under a microscope it looks just like a live plant.

Now let's jump ahead to the end of the composting process. All of the familiar structures of the plant are gone. The cells have all been broken apart. A lot of the large molecules have been decomposed. Proteins are now small amino acids and some of these have even been converted to free nitrate molecules that plants can use. Large starch molecules are now simple sugars. Plants don't use these sugars, but microbes use them to grow, and in the process release nutrients that plants can use. In short, composting converts all of the large molecules into small molecules that can be used by plants and microbes.

Composting is all about creating an ideal environment for microbes. Supply them with the right temperature, nutrients, moisture, and oxygen and they will do all the work for you. For full details of the process see my book *Compost Science for Gardeners*.

Nutrient Cycling

Nutrient cycling is the process that makes nutrients available to all life forms. Small and large animals get their nutrients by digesting food, but what about organisms that can't eat, such as plants and most microbes? They can only feed on free-floating nutrients, and the decomposition of dead organic matter is the main source of these.

All living organisms require a range of nutrients to survive, but the two most critical nutrients for all life are carbon and nitrogen. Earth has lots of carbon in the form of CO_2, so plant growth is not limited by carbon. However, in order for plants to use CO_2 they need a matching amount of nitrogen, and since nitrogen levels are low, it is nitrogen that limits plant growth. Nitrogen is also the growth limiting nutrient for most animals.

The air we breathe is 80 percent nitrogen gas, but it needs to be converted into other chemical forms before plants or animals can use it. These other forms of nitrogen are made available in soil in a number of ways:

- decomposition of special minerals
- lightning
- decomposition of organic matter
- addition of fertilizer
- nitrogen fixation

Lightning converts atmospheric nitrogen into nitrogen oxides (NO and NO_2), which can then be converted into ammonium and nitrate ions by microbes. This is an important process, but it is not a significant source of soil nitrogen.

Fertilizer plays an important role in agriculture, but it is of limited value for native plants.

Organic matter is made in a couple of different ways. As organisms die, their bodies become part of soil organic matter and since microbes do not live very long, they contribute a lot of this material. Organisms that eat others produce some type of waste material—animal and microbe poop—and this adds more organic matter. It all goes through a composting process to release nutrients.

A decomposed one-hundred-pound sample of organic matter produces the following:

- 60–80 lb. carbon dioxide, which gets released into the atmosphere
- 3–8 lb. living microbes

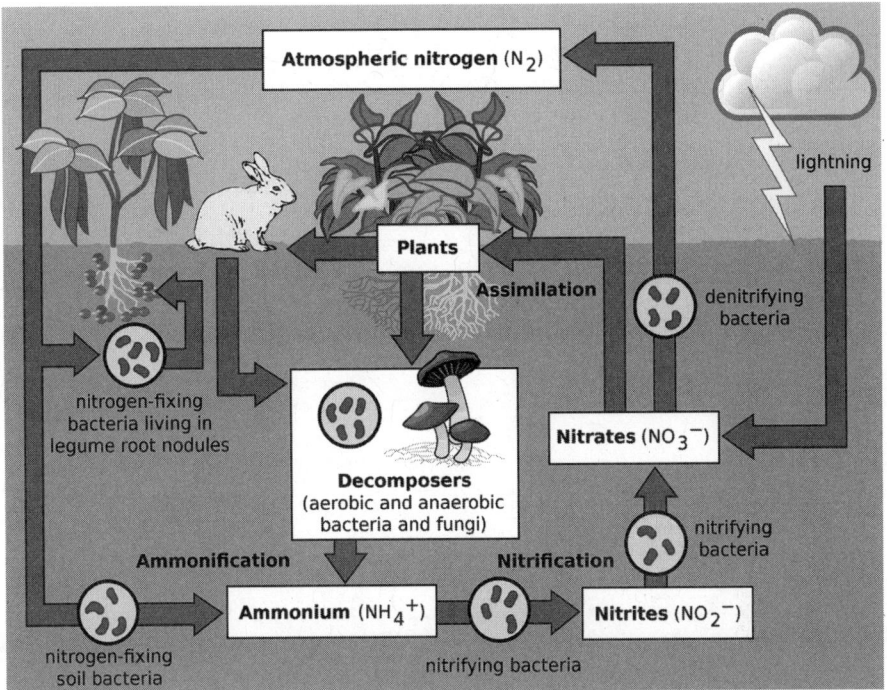

Nitrogen Cycle. Credit: commons.wikimedia.org/wiki/File:Nitrogen_Cycle_2.svg

- 3–8 lb. dead microbes
- 10–30 lb. humus

Organic matter is mostly carbon and oxygen with some hydrogen and nitrogen and very small amounts of phosphorus and sulfur. The amount of other nutrients is even smaller.

Carbon-to-Nitrogen Ratio of Microbes

If you have done some composting, you will know that the right carbon-to-nitrogen (C:N) ratio speeds up the process. The reason for this is that the bacteria and fungi doing the composting need a particular C:N ratio to live, and this is true of all life forms.

The ideal starting C:N ratio for compost is 30:1, thirty times as much carbon as nitrogen. As CO_2 is released during the composting process the ratio falls to 20:1, which is ideal for the microbes.

Each type of organism has a required internal C:N ratio based on their biochemistry. Bacteria tend to be around 5:1 (range of 3:1 to 10:1). Fungi are around 10:1, with protozoa and nematodes having a ratio of 8:1.

The fact that bacteria have a low carbon-to-nitrogen ratio is critical to nutrient cycling and plant growth. When protozoa and nematodes eat bacteria, they consume too much nitrogen relative to the carbon. The excess is excreted as ammonium and nitrate, which can be used by plants and other bacteria.

Building Healthy Soil

As a gardener, you probably dream about black gold—that really nice black crumbly soil. It is great garden soil because it drains well, contains lots of air, and it has many larger channels that are

Microbe Myth: Sterile Soil Exists

Lots of people talk about sterile soil either purchased in a bag or heat treated in the oven. Sellers even put the word "sterile" on their product. The reality is that gardeners never have sterile soil.

Bagged soil may be heat treated to kill pathogens and weed seeds, but the product is not packed using sterile conditions.

You can try to sterilize soil in an oven, but you need a temperature of 212°F (100°C) for thirty minutes to do that. Some people try to use temperatures higher than this, but that can cause soil to become phytotoxic due to soluble salts, such as manganese.

Once soil is sterile, it is impossible for a gardener to keep it sterile. As soon as you open the bag or take it out of the oven, microbes settle on the soil. The plant pot you use is not sterile, nor are your hands or the plant you are potting up.

Is it important to have sterile soil? It will become clear from this book that plants grow better in nonsterile soil. There is no point in trying to sterilize soil, except maybe to kill weed seeds, and that is not a big issue for potted plants.[3]

perfect for root growth. This type of soil is due to a process known as aggregation where sand, silt, and clay are cemented together into larger clumps known as aggregates. This process is described fully in *Soil Science for Gardeners*.

What causes soil aggregation? There are a number of factors, but microbes play a key role. They produce chemical slimes that help stick soil particles together. Soil with more microbes has better aggregation and better nutrient cycling. Microbes are the key to healthy soil.

Microbes Can Harm Plants

So far, I have painted a very rosy picture about microbes and how they help plants, but as all gardeners know, they can also cause diseases that harm plants.

Plants are stationary and are prone to any pathogen that happens to land on them or bump into their root system. They do have an innate immune system that helps defend against some attacks, but this is not their entire defense mechanism.

A big part of their defense is provided by the beneficial and neutral microbes that cover them. These form a protective shield around the plant, both above and below ground. Some of these microbes are known to antagonize pathogens, and others produce substances that are toxic to specific pathogens. Most of these microbes are just taking up valuable space and resources so that pathogens can't gain access to the plant.

From a gardener's perspective, it might appear as if healthy plants suddenly get a disease. Roses get black spot or a lilac gets powdery mildew. And who can forget the late blight that killed all of your tomatoes last year. The reality is that these pathogens have been on the plant long before you saw any symptoms.

Initially, a pathogen invasion doesn't do much damage and the plant can deal with it. You won't even see a problem. In some cases, a point is reached where the fight between plant and pathogen swings in favor of the pathogen. It then grows and multiplies and symptoms start to appear.

Microbe Myth:
Don't Compost Infected Plant Material

Gardeners hate diseases and they will do just about anything to prevent them. Common advice tells you not to compost infected plant material. Instead, remove leaves that show any powdery mildew or black spots and discard them in the garbage. Maple trees get a common disease called tar spots (round black spots), and gardeners are warned not to compost these leaves for fear of spreading the disease.

This sounds like good advice, but in most cases, it's not.

As you learn more about microbes, you will realize that they are everywhere and they travel long distances using air currents. By the time you see white mildew or black spots, the pathogen is everywhere. It's on all the green healthy leaves, the bark, and the ground. Removing a few leaves that show the symptoms will make no difference.

There is also no issue with composting these leaves. The fungal spores are already all over the compost pile and the garden soil so you might as well get some value from the leaves. By fall, every leaf in the garden has some kind of pathogen growing on them. Even discarding every single leaf will have a limited effect on disease next year. Unless everyone in town collects all the leaves with tar spots on them and discards them properly, you will have tar spots next year. They are unsightly but do little harm to the plant.

There are exceptions to this rule, and that is the reason it is important to understand the disease you are dealing with. Viruses are very deadly, and infected plants should be removed and put in the garbage.

Black knot fungus affects mostly Prunus plants (plums, cherries, etc.) and forms black growths on the branches. These growths release spores in early spring. If you remove and discard the growth before then, you will have fewer outbreaks of the disease. The wood from some very contagious diseases such as verticillium wilt in trees should also be discarded.

The key is to identify the problem, understand the problem, and then take appropriate action.

Powdery mildew is a good example of this. Its spores are everywhere, even early in the season. It does very little to the plant until conditions are right, and then it starts to grow. Gardeners usually don't spot it until it is quite far along its development cycle. It is important for gardeners to understand this when they are thinking about treatments and especially preventative options. By the time you see it, it is probably too late.

Gardeners Affect Microbes

A lot of what microbes do happens without much involvement on your part, but you can have a direct impact on them. You can make them happy and prosper, or you can make their life difficult and even kill them. Here are some dos and don'ts for the garden.

Compaction

Most microbes have a similar biochemistry to that of animals and in particular they need to have access to oxygen. The oxygen in soil is supplied by the air that fills the channels between soil particles. Ideal soil has about 25 percent air, which provides plenty of oxygen and more importantly provides a way to dilute the buildup of CO_2.

Compaction reduces the size of the air channels, squeezes air out, reduces oxygen levels, and increases CO_2 levels.

Tilling

Small microbes such as bacteria are not affected very much by tilling, but the larger fungal hyphae are physically damaged.

There are also chemical effects. Tilling introduces more oxygen into the upper soil layer, which in turn speeds up the decomposition of organic matter. In the past this was considered to be a negative effect on soil health, but recent studies have shown that although carbon levels do drop in the top six inches of soil, they actually increase below that level. Tilling does not change the total amount of organic matter in soil. Since most soil microbes live in the top few inches of soil, tilling does decrease populations in the topsoil.

Spraying Pesticides

Spraying pesticides on soil can harm the microbe community, although it is important to understand that every chemical will have a different effect. It is quite incorrect to lump all synthetic chemicals into one group and say they all harm microbes.

For example, glyphosate, the active ingredient in Roundup, has very little effect. When sprayed on soil, it sticks very tightly and is mostly unavailable. We now know that certain microbes use glyphosate as a food source, and decompose it into benign compounds. That explains its short half-life in soil.

Other pesticides persist in soil and we don't really understand how many of these affect microbe populations, but it is almost certain that some will be harmful.

What happens when you spray a pesticide onto leaves? Leaves are covered with microbes that protect the plant from pathogens. Does the pesticide harm these communities? We actually know very little about this for most products.

A few studies have looked at this, and in general it seems that most chemicals will disrupt the microbe community on leaves. Some are harmed, but others seem to prosper. What is clear is that the natural community is affected and any such change may lead to increased disease.

So far I have been talking about synthetic pesticides, but everything I have said also applies to household products. When you spray your dish soap solution on leaves, it appears to do very little damage, but that is because you can't see the microbes and the changes don't happen right away. Soap or any other so-called safe products will affect the microbe community and can lead to a higher level of disease.

Soil Moisture

Almost all of the soil microbes in this book live in the soil water. They need moisture to stay hydrated, and many travel along with the water as it moves through the soil. As the soil dries out, they either die or go into a type of hibernation state known as dormancy. Maintaining a good moisture level is critical for keeping microbial populations healthy.

Microbe Food

Microbes need to eat, and their favorite food is organic matter. They eat each other and they eat dead and decomposing matter. Some can engulf whole organisms, but most just excrete enzymes that digest nearby material so it can be absorbed. Added mulch, an annual layer of compost, and even fall leaves provide food for your microbes.

Some gardeners think they can increase microbe populations by adding "magical" products such as molasses, which is mostly sugar that microbes love. The problem is that it does not last very long. When molasses is added to soil, the food resource suddenly increases, which causes a population explosion. Soon the sugar runs out and microbes start dying. The dead become food for the living, at least for a while. Soon the food runs out and the population returns back to the level that existed before the molasses was added.

This does provide some minor benefit to soil, but it's quite small. It is much better to add other forms of organic matter that take months or even years to decompose and the best of these is

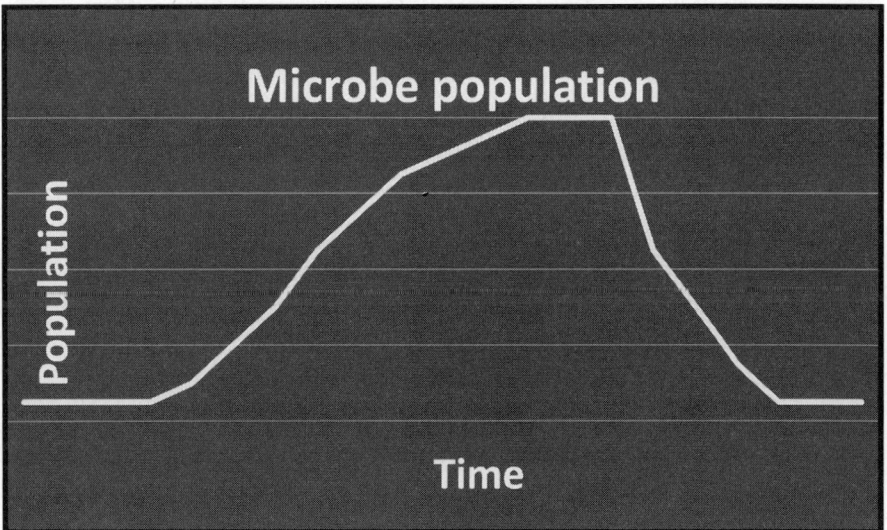

Change in microbe population after adding molasses or other sugar sources.

compost. Which compost is best? This video will explain that to you: https://youtu.be/aONjPeJ-2vM

How Do Microbes Move Around?

Some microbes have appendages that can drive them forward, but most rely on other forces for this such as moving water, air currents, insects, and even gardeners moving soil and plants around. Many fungi form small, lightweight spores that travel great distances on air currents. The molds and powdery mildews can travel many miles to get to your garden.

A lot of pathogens use insects as carriers. Aphids carry all kinds of diseases from plant to plant. Malaria is caused by microorganisms that belong to the genus *Plasmodium* and are moved from host to host by mosquitoes. Many microbes are sticky and become stuck to anything larger that moves: mammals, birds, insects, and even earthworms.

You might think that earthworms like eating organic matter, but the real reason they chomp down on leaves and other debris is to ingest microbes—their favorite food. Microbes also live in their gut and help with the digestion process. What comes out of the end of the worm is a pile of partially digested organic matter and a whole lot of microbes that have just been transported from one place to another.

Some bacteria have flagella, which are a kind of propeller that spins around and moves the organism forward. Others use whips and hairs to propel them.

How do they know where they are going? Some just move in random directions, hoping they go the right way, but many follow chemical signals similar to the way a dog follows a scent.

Bacteria don't have a real body odor, but they release metabolites that do have a scent, and they have a sense of smell. They can detect the smell of ammonia, an important source of nitrogen. This sense of smell has also been detected in fungi and slime molds.

Bacteria can sense nutrients, and will move toward them. Fungi are attracted to chemicals produced by plant roots. Pathogens can "smell" their prey. Movement can be quite deliberate either toward a scent or away from it.

How Much Do We Really Know?

This book is full of information about microbes and it might seem as if we know a lot about them, but we really don't. For instance, we don't even know how many species there are. Scientists think they have identified about 20 percent of the soil microbes. That means we know nothing about 80 percent of them.

We have studied nitrogen-fixing bacteria in root nodules for centuries, but it is only recently that scientists discovered nitrogen-fixing bacteria right inside plant roots. We now know that microbes live inside every part of the plant including leaves, stems, roots, and even inside seeds. These are fairly recent discoveries.

Most of the work has been with individual organisms in isolated containers. A good-quality soil can have thirty thousand species in a teaspoon. We know almost nothing about how each of those species interact with each other, or how the society functions as a whole.

One of the main problems studying soil bacteria is that only about 1 percent can be grown in the lab. If you can't grow them, it is really hard to find out much about them since you have no way to test them. The study of microbes on leaves has been restricted mostly to the ones that grow aerobically for the same reason.

Most plant microbiome studies target only bacteria and fungi and ignore the many other types of organisms. Such studies look at either the above-ground plant or the below ground-plant, but we are learning that the two microbiomes are much more connected than we suspected.

Plants send out chemical signals to attract nitrogen-fixing bacteria so that they form nodules on plant roots. They also send out signals to attract mycorrhizal fungi to their roots. That is fascinating, but even more exciting is the fact that all of the different organisms are sending signals to each other, and we know almost nothing about this.

Don't misinterpret my words here. This is not communication in the sense of animal communication, as so many claim. It is more like yelling to yourself in a dark room, not knowing if any other animal actually exists in the world.

Lots of studies measure the microbial biomass and report it as if this is a number that represents the weight of microbes. The reality is that we don't have a way to measure biomass directly. The tests used for this are indirect measurements that are good in some soil types but don't work in others. Different types of tests give different results. But it is the best we can do right now.

It is important to understand that the information in this book represents a small segment of what we know about plant microbes and that the science on this is certain to change as scientists uncover more and more secrets.

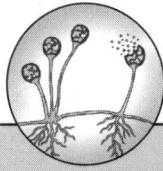

CHAPTER 3

Bacteria

BACTERIA COVER THE PLANET, and are found in every ecosystem. They live on plants, are attached to roots, and are even found inside plant cells. If all the bacteria die, your plants would probably die too; they are that important. A good understanding of bacteria provides insight into why many gardening techniques work or don't work.

Bacteria can't be seen with the naked eye. Five hundred thousand take up no more room than the period at the end of this sentence. A single teaspoon of soil contains as many as thirty thousand species. The sheer number of bacteria in soil is astounding. A gram of fertile soil, about the weight of a paperclip, contains up to a billion bacteria. It is hard to get your head around such a big number but consider this: there are as many bacteria in two teaspoons of fertile soil as people on earth.

There is one certain rule about bacteria and that is, there are no rules. Although most are small, a newly discovered one called "sulfur pearl" (*Thiomargarita magnifica*) is huge and is visible to the naked eye. It is a 2 cm long single thread-like cell that is five thousand times larger than most species.

We tend to think of bacteria as being one type of organism, but the diversity is enormous. There are species that live in virtually every type of environment on earth, and eat almost everything, even spilt oil and jet fuel. Some like it hot, some like it cold, some wet, some dry. Each microbiome hosts a different community of species.

They are single-celled and have a variety of shapes including spheres, rods, and spirals. Most are heterotrophs, which are organisms that get their energy source from carbon. A few are autotrophs and get their energy from soil minerals.

Some bacteria are aerobic and use oxygen in the same way we do, and others are anaerobic. There is even a group that can live in both environments, the facultative anaerobes. An important example of this latter group is *E. coli*, a bacterium that lives in our intestine that is anaerobic, as well as in soil, which is usually aerobic.

For their size, they can move quite far, as much as five micrometers in a lifetime. A human hair has a thickness of about seventy micrometers.

Bacteria can be beneficial, detrimental, or neutral to plants; luckily, most are not pathogens. A few do cause serious diseases with symptoms that include galls, overgrowths, wilts, leaf spots, soft rots, and scabs.

Some plants have immunity to bacterial pathogens, but most of the protection is due to microbial competition. The enzymes they excrete digest both dead and living organic matter, including neighboring bacteria. They also produce toxic substances to kill other bacteria. Since the majority of bacteria are good guys, they tend to win these wars due to sheer numbers. Pathogens have a hard time getting a foothold on plants.

Since bacteria are small, replicate quickly, and are relatively immobile, they are an ideal food for larger organisms such as protozoa, nematodes, and earthworms.

How Do Bacteria Eat?

Bacteria don't have mouth parts, so they don't really eat, but they are able to get nutrients. They have an outer cell wall that separates exterior chemicals from those inside their bodies, and they use something called active transport to selectively move molecules through this wall. It functions similar to the border around a country. There are controls in place that keep unwanted people out while at the same time letting others in. A bacterium wall works the same way by allowing nutrients in and keeping toxic molecules out.

This still leaves bacteria with a problem. Assume they are sitting next to some fresh plant material. All of the nutrients in this material are tied up in large molecules that are too large to transport across the cell wall. Instead of just waiting for something to happen, bacteria take aggressive action. They excrete a variety of enzymes through their cell wall. These enzymes start decomposing the organic matter, breaking the large molecules into smaller and smaller molecules until the nutrient ions are released. The bacterium is then able to absorb these nutrients.

I have described the process in simple terms, but it is actually quite complex. The bacterium does not have eyes or a brain, so it doesn't know if there is organic matter nearby. It also doesn't know which enzymes to produce. To solve these problems, bacteria excrete a wide range of enzymes, hoping that some organic matter exists nearby and that it contains the right kind of large molecules for the excreted enzymes.

Once a nutrient molecule is released, it just sits there, or flows along with the water. The poor bacterium has to wait until the nutrient molecule hits its cell wall and then it can grab it and pull it into the cell. The process seems very inefficient, but it must work reasonably well because bacteria thrive everywhere.

All of this activity—making enzymes, transporting molecules through cell walls, and dividing to reproduce—require a lot of energy. That energy comes from carbon-containing molecules, such as sugar, that the bacterium also absorbs from its environment. This food source is as important as other nutrients.

Organic matter contains a lot of molecules containing carbon, and these are released in the decomposition process. Almost 50 percent of a plant is made up of cellulose, and bacterial enzymes can break this down into sugars that they can use. Lignin is a major component in wood, and it too contains a lot of carbon. Bacteria can't digest it, but fungi can.

Where Do Bacteria Live?

Bacteria are found just about everywhere. In soil, they are found mostly in the top four to six inches (10–15 cm) where there is lots of air, water, and food. The numbers fall off dramatically as you go deeper.

We say they live everywhere, but it is more correct to say that they live in water because the water is necessary to bring food molecules to their cell walls. When soil or the surface of a leaf gets dry they go into a state of dormancy called biostasis, and they wait until water is available again.

Bacteria are coated with a biofilm made of sugars, proteins, enzymes, and DNA that protects them from drying out. This layer holds on to water molecules, slowing down the drying process and protecting them from desiccation.

Bacteria are also very sensitive to temperature, and each species has a preferred temperature range. For this reason, a different community of bacteria are active in fall and winter than in summer. The type of bacteria also changes as a compost pile heats up. Things start off with psychrophilic organisms that prefer to stay below 50°F (10°C). As things heat up, the mesophilic organisms take over (50°F–105°F (10°C–40°C)) and in a really hot pile you find the thermophilic organisms that like temperatures over 105°F (40°C). When temperatures are unsuitable for a particular type of bacteria, they go into biostasis and wait until conditions improve.

These changes in population happen all of the time and gardeners don't really have to worry about them. It is, however, important to understand that there are active bacteria all of the time.

Life Cycle of Bacteria

When it is time to reproduce, bacteria split into two cells, each being a copy of the original. In perfect conditions this can happen every twenty minutes for some species, producing huge amounts in a short period of time. In a lab situation, a single bacterium of *E. coli* can grow to five billion in twelve hours.

Bacterial doubling time is between twenty minutes and several days depending on available resources and on the species. They enter a lag phase when conditions are not favorable, where they just sit around and don't do very much. The number of cells does not change during this phase, but cells do increase in size and are

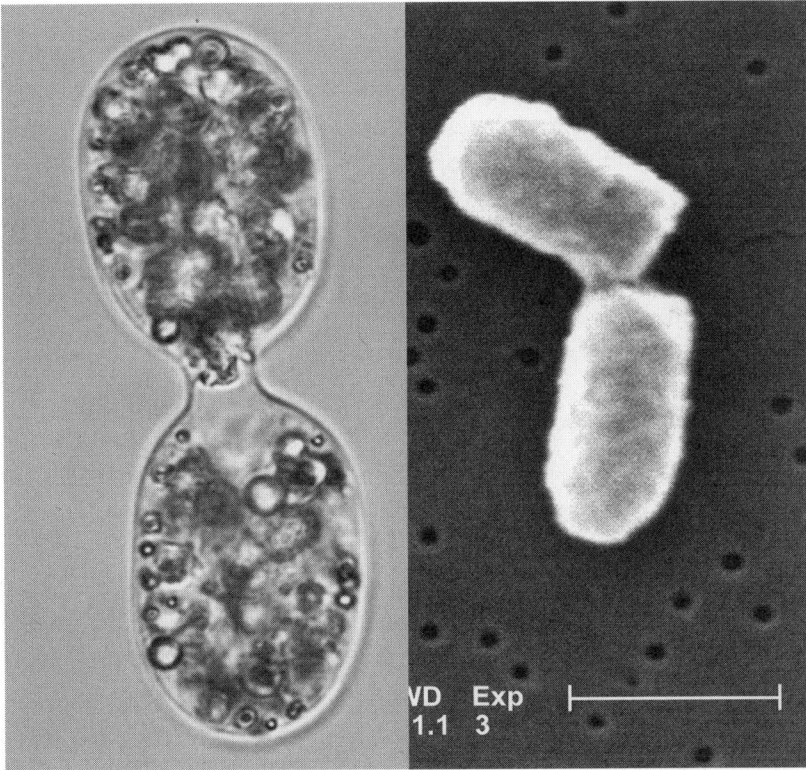

Bacterium on left is starting to divide, the one on the right is almost completely separated. Credit: Djpmapleferryman, https://commons.wikimedia.org/wiki/File:Bacteria_achromatium_Maple_Ferryman.jpg and Janice Haney Carr

metabolically active. When conditions improve, they enter a log phase where they start reproducing at exponential rates.

Bacteria don't have a fixed lifespan because they don't get old. They grow until they divide and form two individuals. These are not really children since they are identical to the original cell. Bacteria do die if conditions get severe, but most deaths are due to predation.

They can also turn themselves into spores, which are a very stable form of dormancy that keeps them alive for a long time in very hostile environments. Bacterial spores have been revived from 250-million-year-old salt crystals.

Nitrogen-Fixing Bacteria

Nitrogen in soil comes from three main sources: dead organic matter, fertilizer, and special nitrogen-fixing bacteria (and archaea) that are able to take nitrogen gas and turn it into ammonium and nitrate ions. This conversion is accomplished using an enzyme called nitrogenase. Oxygen inhibits this enzyme, so the chemical reaction needs to take place in anaerobic conditions. Bacteria have found a number of novel ways to do this.

These bacteria can be classified into three groups based on the type of association they have with plants.

Free-Living Nitrogen-Fixing Bacteria

These bacteria live in soil and are not directly associated with plants. Most of them need a carbon source for energy, which is usually supplied by dead organic matter or plant exudates. A few types can use rock minerals as an energy source.

Most of these bacteria live in anaerobic conditions and because of this it has been historically believed they don't contribute a lot of nitrogen to soil, but recent studies looking at wheat fields in Australia found that they produce 30–50 percent of the plant's nitrogen needs.

This type of bacteria has also now been found inside plant roots. This is an environment that is low in oxygen, high in sugars, and has a low pH—a perfect environment for these bacteria. It has been suggested that these endophytic free-living nitrogen-fixing bacteria might provide most of the nitrogen required by crops such as sugarcane, rice, corn, wheat, and canola.[4]

Associative Nitrogen-Fixing Bacteria

A number of nitrogen-fixing bacteria form loose associations with plant roots and provide a significant amount of nitrogen for cereal crops. The amount of nitrogen depends on several factors including soil temperature, the ability of the host plant to provide a rhizosphere environment low in oxygen, the available exudates, and the competitiveness of the bacteria.

Nodule-Forming Nitrogen-Fixing Bacteria

A specialized group of bacteria form symbiotic relationships with certain plants. The best known of these is the association of rhizobium bacteria with legumes, such as clover, alfalfa, soybeans, broad beans, and peas.

The plant initiates contact by releasing flavonoid-signaling compounds through their roots. The bacteria sense these compounds, and attach themselves to specialized root hairs. The legume then forms a growth, called a nodule, around the bacteria. This nodule not only protects the bacteria, but it also provides the anaerobic environment that is required for nitrogen fixation.

The plant feeds the bacteria by excreting sugars and other nutrients into the nodules. The happy bacteria then fix nitrogen to form ammonia, which is converted to nitrate as it's absorbed into the plant. You can think of these nodules as being little factories that make nitrogen fertilizer for the plant.

Nitrogen-fixing nodules on wisteria roots. Credit: Rowan Adams, https://commons. wikimedia.org/wiki/File:Soil_fertility_-_nitrogen_fixation_by_root_nodules_on_Wistaria_ roots,_with_hazelnut_to_show_size.JPG

This ability of legumes to have their own nitrogen source means that they are able to grow in environments that have very low natural nitrogen levels, making them very competitive. But there is a cost to the plant for this nitrogen. The plant may use as much as 20 percent of its photosynthates (sugars produced by photosynthesis) to maintain the bacteria.

Microbe Myth:
Legumes Such as Clover Add Nitrogen to Soil

Legumes such as clover, peas, and beans use rhizobium bacteria in nodules to fix nitrogen. Gardeners assume that this nitrogen leaks out into the soil and will help feed neighboring plants, but that is mostly a myth.

Think about this from the plant's perspective. It produces a lot of photosynthates and provides them to the rhizobium bacteria. This requires a lot of energy from the plant, but it is worth it to get the nitrogen. Why would the plant now let this nitrogen leach into the soil to feed other plants? It wouldn't. Legumes keep as much of this nitrogen as they can for their own use.

Before a green legume flowers, 60 percent of the fixed nitrogen is found above ground in leaves and stems and 40 percent is below ground. The same plant with mature pods has 80 percent of the plant's fixed nitrogen in the seed, 9 percent in leaves and stems, and the remainder in the roots.

The plant's goal is to make seed, and once pollinated it shunts most of the nitrogen into the developing seed.

Plants also release some nitrogen as exudates and pass some along to mycorrhizal fungi, but most of it remains in the plant.

Legumes are used as cover crops and they do add nitrogen to the soil after they are incorporated into the soil. The best time to do that is when fruit is fully grown. Clover in a lawn adds some nitrogen to the grass if the clover is mowed and the trimmings are left on the lawn. Clover adds very little nitrogen if the trimmings are removed from the lawn.

About half of the nitrogen accumulated by a legume is fixed nitrogen. It is more in low nitrogen soil and less in soil with higher levels of nitrogen. Adding fertilizer reduces the need for the plant to support its symbionts.

There are numerous species of *Rhizobium*, each specializing for a type of legume. For example, peas and beans are infected by different species. In order for the legume to form nodules, the bacteria must be present in the soil. If your soil does not contain the right strain, no nodules will be formed.

Gardeners solve this problem by inoculating seed with the right bacteria before planting. You can either buy little packs of bacteria from seed companies or buy precoated seed.

Once the bacteria is in the soil, it will survive there for several years, so even a four-year crop rotation does not need to be inoculated each time. In some cases, the bacteria survive more than forty years in the absence of a matching legume. Survival is better at higher pH and on soil that is not too sandy.

How do you know if you have the right bacteria in the soil? Grow the legume and have a look at the roots halfway through the summer or in early fall. You can easily see the pea-size nodules if they are there, and they are most visible as plants bloom.

If the plant did not make nodules, you either do not have the right bacteria in the soil or you have too much nitrogen. Excess fertilizer will prevent the formation of the nodules since the plant simply does not need the bacteria.

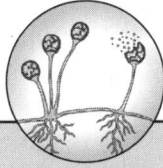

CHAPTER 4

Fungi

FUNGI ARE AS IMPORTANT as bacteria when it comes to plant health. They are decomposers that carry out functions that bacteria are unable to do. Specialized mycorrhizal fungi form symbiotic relationships with roots, helping them get vital nutrients and water.

Fungi are a strange group of organisms. They have some plant-like characteristics, such as the hyphae that look and function similar to roots, and grow out into soil looking for nutrients. But they have no chlorophyll, so they can't make their own food. Instead, they depend on finding carbon compounds for their energy, just like animals. When conditions are right, some form a fruiting body (a mushroom) that functions similar to a flower.

Beer, wine, and bread are produced by single-celled yeast fungi, and the mold on old bread and cheese are the fruiting bodies of fungi.

Living fungi form long hair-like chains of cells called hyphae. The tip of the hyphae branches to form new arms much like the branching on roots. As growth continues, these hyphae form a large, fuzzy-looking mass called mycelium. You can't see individual hyphae since they are much thinner than a human hair, but you can see the mycelium. The hairy green circles on an old orange is a good example.

A single gram of soil, the weight of a paperclip, can contain three hundred to three thousand feet (one hundred to one thousand meters) of hyphae. Fungi are less numerous than bacteria, but since each organism is larger, the total mass of fungi in soil can be two or three times the mass of bacteria and together they account for most of the microbe population.

Over one hundred thousand species of fungi have been identified, of which seventy thousand are found in soil. It is estimated that the total number of species is around 1.5 million. We still have a lot to learn about fungi.

When they die, they leave behind a vast network of tunnels in the soil that allow water and air to move through, as well as other smaller organisms.

Fungi are able to condition their surrounding environment. They exude a wide range of chemicals, including acids, to adjust pH and antibiotics to help control parasitic organisms. Penicillin is probably the best known fungal antibiotic.

In addition to whole organisms, fungi exist as "propagules," a term used to refer to a bit of fungi that is able to grow into a fully functional organism. Propagules include spores, pieces of hyphae (filaments), and sections of colonized roots. Many commercial products quantify the fungal microbes in their product by listing the amount of propagules on the label.

You can think of propagules as seeds ready to sprout and grow as soon as conditions are suitable, similar to the plant seed bank in soil. A soil that has a naturally high inoculum potential (i.e., can grow a lot of fungi) has around fourteen propagules per gram of soil.

Yeast are a special type of fungi that is treated in a separate section in this book.

What Do Fungi Eat?

Fungi are heterotrophs, which means they cannot make their own food. They need to find a carbon source, and they do this in a very similar way to bacteria. They excrete various enzymes and other compounds that decompose the large organic molecules that surround them. Once converted to small nutrient and sugar molecules, they can be absorbed and transferred several feet along the hyphae to the point where they are needed.

Unlike other microbes, some fungi can digest really tough organic matter such as lignin and cellulose (paper). The tip of the hyphae produces enzymes that let it penetrate hard surfaces such as plant leaves and bits of old stems to get at the more digestible material inside. Most of the absorption of food takes place near the tip of the hyphae.

A unique quality of fungi is their ability to grow hyphae above the soil line to penetrate leaves and other plant debris lying on the surface. They are critical for cleaning up the plant litter that drops in fall. The nutrients in this material are then moved deeper into the soil and will eventually be used by plants and other microbes.

This above-ground growth is very evident in a pile of leaves. The leaves don't contain enough nitrogen for bacteria to decompose

(A) fungal hyphae on wood chips, (B) dog vomit slime mold (*Fuligo septica*), (C) octopus stinkhorn (*Clathrus archeri*), (D) dead man's fingers (*Xylaria polymorpha*). Credit: Geo Lightspeed7, Ryan Hodnett, Public Domain[5]

them, but fungi soon take on that role and produce leaf mold. Leaf mold is a combination of partially decomposed leaves and lots of visible mycelium, making it a great addition to any garden.

Wood chip mulch is also invaded by fungi. After several weeks, the wood pieces are stitched together with fungi hyphae that slowly decompose the wood. Some of these fungi even form harmless fruiting bodies right on the wood. Some of the weirdest looking ones include dog vomit, stinkhorn, and dead man's fingers. These are perfectly harmless and indicate a healthy soil environment.

Where Do Fungi Live?

Fungi are aerobic and unlike most other microbes do not need a water film to survive. They grow in soil, in the air above it, and on most other surfaces including leaves, stems, flowers, and fruits. The white dusting of powdery mildew is produced by a fungus.

Hyphae are thinner than plant roots but thicker than bacteria. They travel through pores in soil that are bigger than the hiding places used by bacteria and smaller than the openings used by roots.

Life Cycle of Fungi

Fungi can reproduce sexually or asexually, which is more common. Asexual reproduction can be carried out in two ways: fragmentation and by the formation of spores (conidia). In both cases, the offspring are genetically identical to the single parent.

Fragmentation

Many fungi can reproduce from pieces of hyphae called fragments, which are a type of propagule. Each fragment can grow into a whole new organism. Fragmentation can happen as a result of damage such as tilling, or it can be initiated by the fungi itself.

Sporulation

Sporulation, the formation of spores, is the most common form of asexual reproduction. Special hyphae develop that then produce vast quantities of spores.

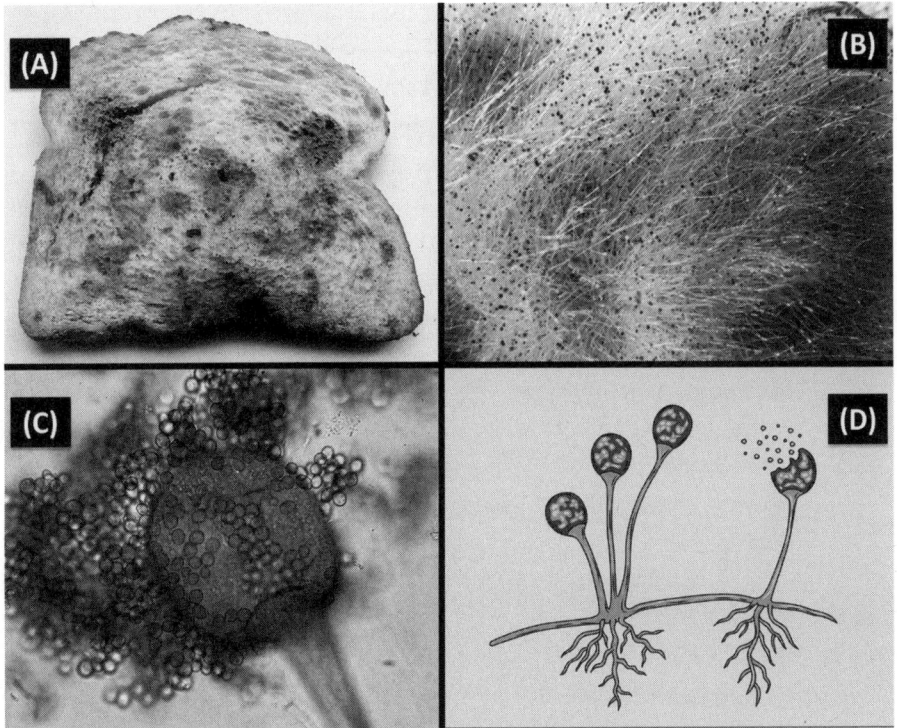

Black bread mold, (A) moldy bread, (B) close-up of *Rhizopus stolonifer,* (C) spores being released, (D) structure of the mold. Credit: Vincent van Zeijst, WDKeeper, George Shepherd[6]

Bread mold looks like a black or blue-green fuzzy surface, but if you have a closer look you will see vertical stems with black spheres on top. These stems are the special hyphae and the round tops are sacs that contain hundreds of spores. When the spores are fully formed, the sac bursts open and releases them.

The mushrooms in your lawn look different, but under a microscope you will see similar hyphae producing spores very much like the bread mold.

Sexual Reproduction

Fungi normally use sexual reproduction only when the environmental conditions are not favorable. This is probably done in the hopes that jiggling the genetic material around will produce off-spring better able to live in the new environment.

The process is complex, and several different methods are used by different species. Some produce differentiated sex organs that release sex cells (gametes). Each sex, designated as "A" and "a," can be produced on different individuals or on the same fungi. In the latter case, the two sexes are incompatible with each other and require a second organism to mate.

Others produce special organs (gametangia) that touch one another to exchange genetic material. More advanced fungi use their normal hyphae and exchange genetic material when they come in contact with each other.

In most cases, sexual reproduction results in the formation of spores. These will be genetically different from the parent producing them.

A Spore's Life

If you have ever stepped on an old puffball mushroom, you have experienced the release of a dust cloud made up of fungal spores. If you take a mushroom cap and let it sit for a day, you'll find powdery spores under it.

Spores are a special kind of seed that are extremely small. A thousand would easily fit on a pinhead. They easily float through the air, and the breath you just took sucked in a bunch of them. They travel great distances on wind currents, which means that fungal spores are everywhere. They exist universally in all climates and can even be found in the Antarctic.

Some spores are spread by water droplets from rain or the spray of streams. Others are moved around by animals and insects. The stinkhorn fungi release their spores into smelly slime that attracts flies. As the flies eat the slime they also consume the spores, which are later deposited through fly poop.

When spores land in a suitable environment, they sprout and start growing hyphae. They can grow as fast as forty micrometers per minute, which is the thickness of a human hair, so it doesn't take long to form a visible clump of mycelium.

Pathogenic Fungi

The majority of fungi are good guys that are beneficial to plants. Unfortunately, there are also some bad actors that parasitize plants. You might know them as mildew, root rot, rust, damping off disease, botrytis, scabs, cankers, and fusarium. Fungi cause more harm to plants than any other soil organisms.

In many cases, they don't kill the host, but they do weaken it. In agricultural crops they reduce the yield, and in horticultural plants they can reduce vigor and flowering. Soil-borne fungi tend to invade roots and form some kind of deformed growth, such as club root on cabbage.

Fighting these diseases needs to be done on an individual basis. Identify the problem and then research specific ways in which to reduce the issue. There is no general cure that works for all fungal problems.

Mycorrhizal Fungi

Mycorrhizal fungi are a special group of fungi that form a very close symbiotic association with plants. The fungi receive sugars and other nitrogen-containing compounds from the plant in exchange for water and mineral nutrients. It is estimated that 95 percent of all plants form such a relationship.

An interesting example of this is the fungal association that is formed with the Indian pipe (*Monotropa uniflora*), also known as the ghost plant or corpse plant. This is a white mushroom-like herbaceous plant that has no chlorophyll. It forms an association with mycorrhizal fungi, which in turn forms an association with trees. The fungus receives sugars from the tree and shunts them to the Indian pipe plant.

The fungal hyphae extend the effective length of the average root system by a factor of one thousand, and increases the surface area sixty-fold, making it much easier for plants to get nutrients. The hyphae are also much thinner than roots, so they are able to collect nutrients from pores that are too small for roots. Fungal enzymes and acids also free nutrients such as copper, calcium, magnesium, and zinc from rock material.

Mycorrhizal fungi connects the Indian pipe to trees as a food source.

Phosphorus levels are always low in the soil solution, but fungi are able to pick up significant amounts due to their large surface area. This is the most important nutrient they provide to plants. It should also be noted that too much fertilizer can increase phosphorus levels in soil, which inhibits mycorrhizal growth.

The benefits of mycorrhizal fungi to plants is significant and includes the following:

- increase yield in agricultural plants from 10 to 40 percent
- remove soil contaminations by absorbing toxins
- reduce fertilizer needs
- increase tolerance for diseases, drought, and chilling

Seedlings are colonized by fungi as soon as their roots contact the mycelium network. There is also some chemical signaling between plants and fungi that lets each know the other is there. A kind of "hello, I'm here—do you want to date" type of signal.

A single root can be colonized by five or six species at the same time. The amount of colonization in crops depends on the crop. For instance, radishes are not colonized at all.

The plant shunts as much as 20 percent of its photosynthates (sugars) to the fungi. This is a significant drain on the plant, and only makes sense if the plant gets enough benefits from the fungi. In nutrient-poor soil the plant has trouble getting enough nutrients and it is quite willing to feed the fungi for more nutrients. Things change in nutrient-rich soils, where it is easier for the plant to get its own nutrients and keep the photosynthates for itself. The plant transfers less food to the fungi, which starves the fungi and results in less fungal growth. This process is under the control of the plant.

These fungi have been divided into two main categories; ectomycorrhiza (outside root cells) and endomycorrhiza (inside root cells).

Ectomycorrhiza

Ectomycorrhiza (EM) are less common, but they are important for woody plants. An individual tree may have associations with fifteen or more different species at the same time. There may be as many as twenty thousand species in soil and they form associations with 5 to 10 percent of all plants. They are important for Christmas tree production and the agronomic crops of hazelnuts and pecans.

EM accounts for 30 percent of the microbial biomass in forest soils. The external mycelium may be more extensive than that of AM fungi (described below) with as much as 650 feet (200 m) of hyphae per gram of dry soil.

Most of the hyphae of these fungi wrap around roots, forming a type of sheath, but some do enter the root and grow inside the intracellular space between root cells. Once the sheath is formed, they are able to exchange compounds with the roots. This group

Mushrooms formed by mycorrhizal fungi, (A) *Cantharellus cibarius,* (B) *Lactarius deliciosus,* (C) *Morchella esculenta,* (D) *Laccaria bicolor.*
Credit: Ivan Teage, Gljivarsko Drustvo, public domain, Mary Smiley [7]

forms many of the noteworthy mushrooms including amanitas, chanterelles, and the prized truffles.

In EM dominated soils, such as coniferous forests, the decomposing organic matter may be a more important source of minerals than the soil itself. The fungi are able to extract nutrients from this organic matter and shunt it to plants. EM fungi can also sequester large quantities of nitrogen in their mycelium. This nitrogen is then moved to host plants when nitrogen demand is high in spring during bud break.

Endomycorrhiza

The endomycorrhiza form associations with 85 percent of all plant species including most horticultural plants and agronomic crops.

The hyphae of these fungi enter the roots and grow directly inside root cells.

This group can be further broken down into the following subcategories:

- Ericoid mycorrhiza, which form connections with the acid-loving plants in the Ericaceae family
- Orchid mycorrhiza, which are required for orchids to grow in natural settings
- Arbuscular mycorrhiza (AM or AMF), which is the most common mycorrhizal fungus

Arbuscular Mycorrhiza (AM)

These fungi get their name from the fact that they penetrate root cells and then form arbuscules—tree-shaped structures—at the end of the hyphae. These structures dramatically increase the surface area of the fungi inside the cell, making exchange of water and nutrients much more efficient. Normal root hairs are about 1–2 mm long and the mycorrhizal fungi extend this reach to about 120 mm.

Fossil evidence indicates that the earliest AM arose 450 million years ago. It is estimated that there are now three hundred thousand species, of which about 240 have been identified. Unlike most fungi, AM have a limited ability to degrade organic matter. They are obligate biotrophs, which means they need living plants to survive. Without them they have no energy source.

The diameter of a hyphae ranges from 2 µm to 20 µm, with root hair being about 10 µm. The total length of hyphae ranges from 1 to 110 m/g of soil. Many factors including soil type, environmental conditions, and plant type affect the amount of fungi material in soil.

A major benefit of AM is the formation of soil aggregates, which increases the size of pores and gives soil better water-holding properties. They do this by excreting a sticky substance called glomalin. Although AM colonization is generally beneficial to plants, there are cases where they decrease plant growth. In some matchups,

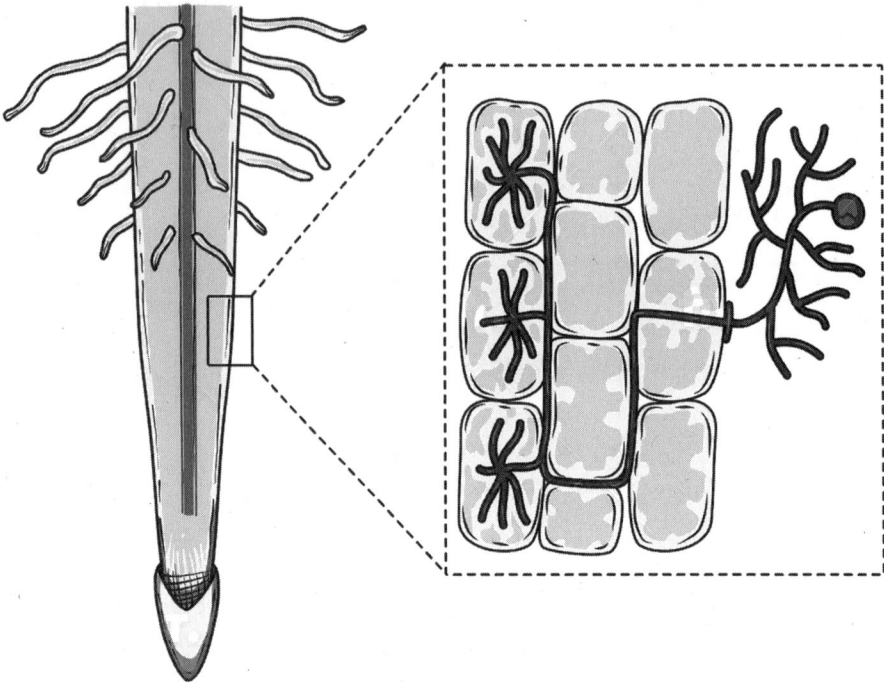

AM infecting a root and forming arbuscules inside root cells.

the plant passes too much photosynthate to the fungi, effectively starving itself.

Altered Plant Growth

The presence of mycorrhizal fungi help plants grow better, but they also have some specific effects on plants and plant ecosystems.

The hormonal interactions between fungi and plants can have dramatic effects on the roots. Plants grow fewer root hairs with a fungal symbiont because they don't have to collect as much water and nutrients themselves. They also tend to grow smaller root systems.

The Big Bluestem grass, *Andropogon gerardii*, grows in both phosphorus-rich soil and phosphorus-poor soil. The group of plants in P-poor soil have a higher tendency to form associations with mycorrhizal fungi and they have smaller root systems because they rely more heavily on fungal assistance. The same species growing in P-rich soil develop larger root systems to collect their own phosphorus.

Such changes in plant growth also influence species composition and diversity in natural environments. Plants that can form strong bonds with fungi can become more competitive and have an easier time finding resources in a crowded space. Mycorrhizal species also vary quite a bit in their ability to grow hyphae and collect resources. Plants that are adapted to use the more successful fungi will dominate an ecosystem.

Global warming due to higher CO_2 levels allows plants to increase the level of photosynthesis, providing them with excess photosynthates. It is expected that plants will make these available to fungi, resulting in an increased fungal biomass as the world heats up, at least in colder climates. In climates that are already hot, we will probably see reduced plant and fungal growth because the temperatures will get too high for them.

The Common Mycorrhizal Network (CMN)

The common mycorrhizal network is also called the "social network" of trees and fungi, even though the network can also exist between non-trees and fungi.

In the above sections I have described how mycorrhizal fungi connect to the roots of a single plant, but what happens when such a fungus also connects to the roots of a second, third, fourth, or even fifth plant? That is the CMN.

Consider two trees, a birch and a pine. Each has their own root system and the two root systems are completely separate from one another. Along comes a mycorrhizal fungus that makes a connection to both trees at the same time. Each tree passes exudates to the fungus and the fungus passes water and nutrients to the trees. Neither tree has any idea that the fungus is double dating.

Fall comes along and the birch drops its leaves and stops making photosynthates. The fungus has just lost one of its carbon sources, but not to worry. The pine is still green and produces exudates to keep the fungi alive.

In spring, the birch starts growing again and has a great need for water and nutrients. The fungus has not been getting any sugars from the birch, but it does have a steady flow from the pine, so it is able to continue pulling resources from soil and passing it along to the birch. It can even take some of the sugars from the pine and pass them along to the birch.

Once the birch has fully leafed out, it will be in full production and it passes excess photosynthates to the fungus, who can pass some of them on to the pine, which tends to photosynthesize at a slower steady pace in summer.

Plants initiate the symbiotic relationship with fungi, but it seems as if the fungi has more control over the marriage than previously thought because it controls the amount and type of compounds passing to and from plants.

All kinds of other chemicals can also be passed from one tree to another. Some people call these "infochemicals" to reflect the fact that many of these act as signaling compounds. People have interpreted this transfer of signals as "communication."

Humans like to anthropomorphize other organisms—I just did it by talking about dating and marriage—and some refer to the transfer of chemicals between organisms as communication, but that is incorrect. Communication implies that both the sender and receiver have knowledge that communication is happening and that both parties understand the message. That does not happen between trees.

The CMN described above has been reported in quite a few studies. Admittedly, there is not great agreement on the specific aspects of the story and the details vary a lot based on plant types and the environment. Most results are from lab tests done in pots, and this data can't necessarily be extrapolated to tell us what actually happens in the field.

The second problem with the story is that it is very difficult to measure the transfer of these compounds through the fungi and to properly adjust for the natural transfer through the soil solution. Remember that plants also excrete photosynthates into the soil, and from there they can be picked up directly by other plant roots without fungi.

A meta study dated 2021 looked at the current research and concluded that movement of chemicals between plants seems to happen, but that there is no solid evidence that the fungi is responsible for the transfer. The results can be explained by direct transfer between plants via the soil. The full story won't be known until more work is done.[8]

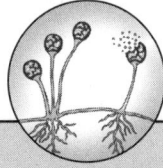

CHAPTER 5

Yeast

MOST OF YOU PROBABLY RECOGNIZE the term yeast as it relates to baking, wine making, and brewing beer, but these organisms are also important to plants. Yeast are found all over the place including plants, animals, insects, compost, water, soil, and they even float in the air. You ingest a bunch of them each time you bite into an apple or eat a grape. They play important roles in nature, including the decomposition of organic matter. Fall leaves and banana peels slowly decompose in part due to yeast.

Yeast are a form of fungi, but unlike most of the fungi described earlier in this book, they have a single cell and their structure is more similar to that of higher organisms including humans. Unlike simple organisms, such as bacteria, their DNA is contained in a nucleus and they have Golgi apparatus, mitochondria, vacuoles, and a thick cell membrane. They can even reproduce in a quasi-sexual manner.

They are mostly benign, but some species are pathogens. Vaginal yeast infections are caused by *Candida albicans*, and a similar species causes diaper rash and thrush of the mouth. Another particularly nasty pathogenic yeast, *Cryptococcus neoformans*, produces life-threatening meningitis.

Most yeast convert carbohydrates (sugars) into CO_2 and alcohol using a process called fermentation. It forms the basis of many of our alcoholic beverages, and it is also what causes bread to rise.

About fifteen hundred species of yeast have been identified, and they range in size from one to five micrometers wide and two to thirty micrometers long. They live alone or as multicellular colonies where they frequently form strings of individuals, not unlike a string of pearls.

Like other fungi, yeast are unable to carry out photosynthesis and therefore rely on other sources for their energy needs. Their main food are simple carbohydrates such as sugar and acetate (vinegar), and as with all life forms they also require a source of nitrogen and other nutrients.

Each species has a preferred temperature range. For example, the yeast used for making bread thrives at about 85°F (30°C), which is why bakers keep their rising dough warm. Too cold and growth slows down. Too hot and it dies.

Yeast is a facultative anaerobe, so it can live in both an aerobic environment and an anaerobic one. Fermentation takes place without oxygen.

Life Cycle of Yeast

For such a small organism, yeast has a complex life cycle, which includes both sexual and asexual reproduction.

In a suitable environment that has enough moisture, food, and is at the right temperature, yeast reproduce asexually through a process called budding. As cells grow, they form a small bud, or protrusion, on the outer cell wall. Once the cell is large enough, it duplicates its DNA and the nucleus divides into two. One nuclei stays in the mother cell, and the other moves into the developing bud. The bud then breaks off to form a new daughter cell. This form of reproduction can occur every ninety minutes in a good environment.

When the environment becomes less hospitable, yeast use a different form of reproduction. The original DNA is duplicated, but instead of forming two complete cells it forms four haploid cells, each containing half of the normal thirty-two chromosomes. Two types of haploid cells are formed, one called the a-factor and the other called the alpha-factor. This process is called sporulation, and the haploid cells are known as ascospores or just spores. Each type of spore produces a different pheromone that can be detected by the other type, and the two types of spores are analogous to the sex cells in higher animals.

The spores can continue to divide asexually producing more haploid cells or they can go into a state of dormancy. Once dormant they are very stable and can survive drying out and even freezing. When the environment improves they become active again.

When conditions are right, a form of sexual reproduction takes place where an a-factor spore combines with an alpha-factor spore to form a complete, fully functional, diploid yeast cell.

The above life cycle describes *Saccharomyces cerevisiae* and gives you a general understanding of how yeast reproduce, but each species does things a little differently.

Yeast-Plant Interactions

Plants are covered in yeast. They live right next to roots in the rhizosphere, they cover leaves and stems, they are found inside flowers, and they even coat the outside of fruit. The above-ground parts of plants tend to have a different composition of species than the below-ground parts.

Yeast and plants clearly have a close association with each other, but these interactions have not been studied as well as plant-bacteria or plant-fungi interactions.

Yeast in Hellebore Nectar

Helleborus foetidus, the stinking hellebore, is one of my favorite garden plants for a number of reasons, but especially because it is evergreen in zone 5 and looks good for twelve months of the year. They bloom quite early, even with snow on the ground. By the way, the "stink" is so mild you have to rub the plant to detect it.

Bumblebees like visiting such early flowers and we now know that they deposit yeast into the nectar of the flower. Nectar is basically sugar water, which is perfect food for yeast growth. As the yeast grows and multiplies they generate heat just like any other organism, and this heat warms the inside of the flower.

The increase in nectar temperature, which can be as much as 2°C above ambient, increases the evaporation of volatile organic compounds, which may help attract pollinators. But there is a downside to this whole process. The yeast uses up sugars to produce the heat, which reduces the amount of sugar in the nectar, making it less appealing to pollinators. The question then becomes, do bees prefer heat over sugar-rich food? They probably prefer heat on cold days and more sugar on warm days.

You might wonder where bees get the yeast. It is found on their bodies and in their stomachs. Numerous yeast species are also found on most flowers. Remember, yeast is everywhere.

Yeast Creates Alcoholics

There is another downside to yeast growing in nectar. As spring turns to summer and things warm up, the yeast in nectar becomes even more active and they do what yeast does best—make alcohol. Bees, especially honey bees, frequently get drunk by sipping too much alcoholic nectar.

A similar thing happens to birds in late fall and winter. Fruit is covered with yeast, and as the fruit gets overripe the yeast starts fermenting the sugars inside. Birds then come along, eat the alcoholic juice, and get drunk. Bats and butterflies also like to get drunk on fermented fruit.

Yeast on Leaves

The phylloplane (surface of leaves) is covered with all kinds of microbes, including yeast. Each species has a preferred environmental condition where it grows best, so the actual species on leaves varies by location, plant type, and time of year.

When scientists looked at yeast on five species of fruit trees, they found 155 strains belonging to 11 genera. Only three species occurred regularly. [9]

Almost no yeast is present on the young spring leaves of the common wood sorrel *Oxalis acetosella*. As the summer and fall progress the number increases, with the highest number being present after snowfall. Species diversity is highest in autumn.

In general, similar species of plants will host similar species of yeast, but there can be wide variations.

Powdery Mildew

Powdery mildew is a common fungal infection on plant leaves. A yeast species *Pseudozyma flocculosa* has been shown to slow down the growth of several species of powdery mildew including those on roses and wheat. Work is now being done to develop this as a potential biocontrol agent.

It is important for the gardener to understand that this is not a simple process of spraying the right yeast so that powdery mildew goes away. The antagonistic properties of the yeast depend very much on the presence of other compounds on the leaf. It may also be plant/fungal species dependent. For example, the epiphytic yeast *Pseudozyma aphidis* controls powdery mildew on cucumbers, but other types of yeast don't work.

Too often gardeners are led to believe that biocontrols are easy to use, but things in the real world are rarely that simple.

Microbe Myth: DIY Yeast Biofertilizers Work

The yeast in soil has some clear benefits for plants, so it should be no surprise that people have developed DIY biofertilizers containing yeast.

A basic mixture of water, sugar (or molasses), and some baker's yeast (*S. cerevisiae*) is allowed to ferment for several days. To this basic recipe people also add compost, manure, milk, wood ash, rock dust, etc. The resulting mixture is then applied to soil or even used as a foliar spray.

Does this work? It is important to define the word "work." We know that manure contains nutrients, and when added to soil it helps plants grow. But the real question that needs to be asked is, does adding baker's yeast to soil or leaves benefit the plant?

Remember that yeast is everywhere, so it is already present in soil, on plant leaves, in compost, and in manure. The second thing to understand is that there are many species of yeast, and gardeners really only have access to baker's yeast. Why would they think this is the right species for all plants?

There is no scientific evidence that these DIY concoctions work. Just apply the manure and compost, and stop making gardening difficult.

Yeast in Soil

Soil yeast produces a variety of biologically active compounds (auxins, phytohormones, vitamins, amino acids, enzymes, etc.) that can stimulate plant growth and development.

Yeast can control pathogenic organisms by improving the plant's access to nutrients which in turn helps the plant grow and fight its own battles as well as produce antimicrobial agents that combat pathogenic bacteria and fungi in the rhizosphere.

The black soil under a chestnut tree contains over five hundred strains of yeast with the greatest number found at a soil depth of four to eight inches (10 to 20 cm). Some of these strains inhibit the growth and development of pathogenic fungi such as fusarium blight.

The number of yeast found in the rhizosphere are much higher than in bulk soil, and as a plant matures and grows larger the number of soil yeast increases. For example, more yeast is found under old apple trees than young trees. Some yeast species have been shown to form a symbiotic relationship with roots, resulting in increased plant growth.

A Bioindicator of Air Quality

Yeast on leaves are sensitive to air pollution, especially sulfur dioxide. Both the number of cells and the number of species drops as the air quality decreases, and this change in population dynamics has been used to quantify air quality.

A Possible Solution to Plastic Pollution

Could yeast solve our plastic pollution problem? Maybe.

It is still early days in the research, but special strains of yeast have been developed that produce an enzyme known as LCC (leaf branch compost cutinase). This enzyme digests certain types of plastic and is more environmentally friendly than current plastic recycling options.

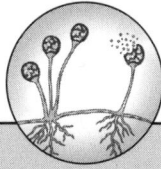

CHAPTER 6

Nematodes

NEMATODES ARE NOT MUCH MORE THAN small wriggling tubes, but they are very successful. It is said that if you were to count all animals on Earth (not counting true microbes), four out of five would be nematodes. They are everywhere: in soil, at the bottom of the ocean, and living as parasites in plants, animals, and humans.

They can swim short distances in water films where they graze on bacteria, fungi, protozoa, and smaller nematodes. Some larger nematodes are available in two commercial products, one for controlling slugs and snails which is not available in North America.

Most nematodes are found in the top six inches (15 cm) of soil where they can consume five thousand bacteria a minute. Like protozoa, they need less nitrogen than the bacteria they eat and consequently excrete the excess into the soil solution for plants to use. They are key contributors to soil nitrogen mineralization. They are up to 2 mm long and 50μm wide, and a teaspoon of soil contains fifty to five hundred of them. There may be as many as forty thousand species worldwide.

The size of nematodes varies a lot, as does the reproductive cycle, which can take as little as three days or as long as three years. On average, they take thirty days to reproduce.

Nematodes are simple multicellular organisms with no circulatory system or respiratory system. They are so thin that they can use their skin to absorb oxygen and excrete waste. They are able to adapt to most environments. In agriculture, they cause significant losses in yield as well as the cost of the fumigation used to kill them in soil.

Nematodes can be beneficial or parasitic to plants. When they attack plants, they puncture roots and suck plant juices out, weakening the plant. These puncture wounds then provide an entrance for other pathogens. Roots respond by making root knots or swellings. A good example of this is the root-knot nematode that affects carrots[10] and the garlic nematode[11] that is destroying huge amounts of crops. OMAFRA (Ontario Ministry of Agriculture, Food, and Rural Affairs) has informed me that most of the seed garlic in Ontario is now infected.

They can also be carriers of viruses. The dagger nematode can infect fruit trees with the tobacco ringspot virus that causes apple union necrosis, as well as the cherry rasp leaf virus that causes rasp leaf disease.

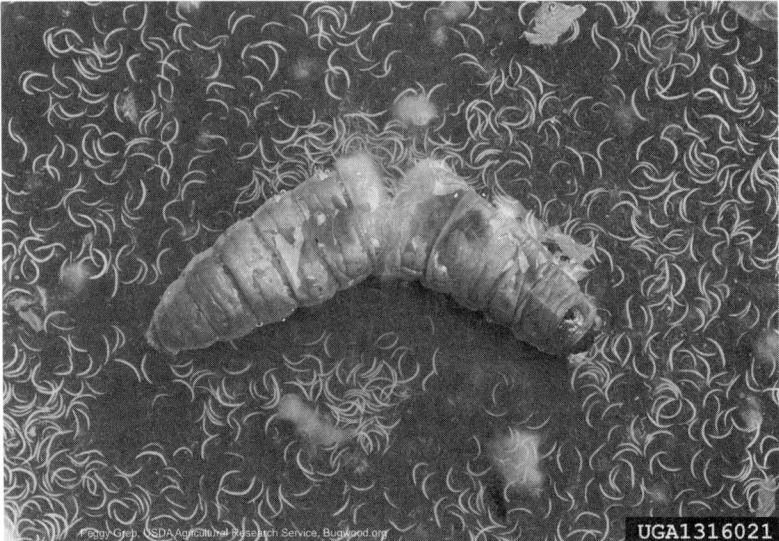

Nematodes escaping from their dead host larvae. Credit: Peggy Greb, USDA Agricultural Research Service, https://commons.wikimedia.org/wiki/ File:Entomopathogenic_nematode_%28Heterorhabditis_bacteriophora_%29_ Poinar,_1975.jpg

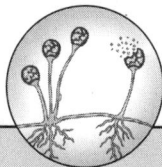

CHAPTER 7

Protozoa

IF YOU THINK BACK—way back to biology class—you might re-member a little organism called an amoeba. It is large enough to be seen with a light microscope, which makes it ideal as a teaching aid. This amoeba looks like a blob, but it's a very interesting proto-zoan that plays an important role in soil.

Protozoa are not technically animals, but it is useful to think of them as such. They are single-celled, and move around the soil solution looking for prey, which consists mostly of bacteria. Size depends on species, ranging from two to several hundred microm-eters. The top couple of inches of soil are a favorite hangout, and they can be found globally in a wide range of climates from warm, dry deserts to cold, damp tundra.

About fifty thousand species have been identified of which about sixteen hundred live in soil. It is expected the total number in soil to be closer to four thousand. Although protozoa are rela-tively small, their combined biomass in healthy soil is greater than all other soil microorganisms combined. The annual production of protozoa on a weight basis is about the same as the production of earthworms, both of which add a significant amount of organic matter to soil as they poop, die, and play a key role in nutrient cycling.

Some have a mouth-like aperture for sucking up prey, but most simply engulf their prey. Once inside, their digestive en-zymes digest the food. The vampire amoeba, called *vampyrellids*, "chews" perfectly round holes in fungi and uses them for food. They do this by attaching to the surface of the fungi and excreting enzymes that digest the hole. The amoeba then sucks up digested fungal cells.

Bacteria grow very rapidly, and they would take over the soil were it not for protozoa who gobble them up as fast as possible. Paramecium, a very large protozoan, can consume more than five million bacteria daily, and even smaller species consume ten thousand a day. The number of protozoa is directly proportional to the number of bacteria, and since the highest population of bacteria is in the rhizosphere, that is also where you find the highest density of protozoa.

Protozoa also eat other foods including other organisms such as algae, fungi, smaller protozoa, and some pathogenic nematodes. Some varieties consume a lot of dead organic matter, while others eat roots and damage plants, but most are not a pest in the garden.

This all sounds great for protozoa and it should result in them overwhelming the soil, but they have predators too. Nematodes, larger protozoa, and arthropods find protozoa very tasty and keep their numbers in check.

Some protozoa contain chloroplasts that photosynthesize to produce sugars.

Unlike bacteria and fungi that have limited mobility, protozoa use specialized appendages to move around. We talk about them living in soil, but in actual fact protozoa are water-dwelling microbes. They live in the soil water that surrounds soil particles and fills soil pores. This makes them prone to droughts. If it gets too dry, they form a cyst and go dormant until moisture returns. Some species can survive up to a year in this condition.

Most of the discussion here describes protozoa in soil, but they also live inside other organisms. Flagellated protozoa such as *Trichonympha* and *Pyrsonympha* inhabit the guts of termites, where they help digest wood. A wide range of protozoa live commensally in the stomachs of ruminant animals such as cattle and sheep. Some ciliates live in a symbiotic relationship inside the gut of annelid worms. All higher animals are infected with one or more species of protozoa. Protozoa also live on plant leaves and stems, but the number found there tends to be much smaller than other microbes. The environment is too dry for most of them.

Protozoa are broken down into three main subgroups based mainly on their mode of locomotion: ciliates, flagellates, and amoebae.

The Amoeba Group

There are two types of amoebae: the naked and the testate.

The naked group are known as "shapeshifters" due to their irregular changing shape. They move by creating foot-like projections

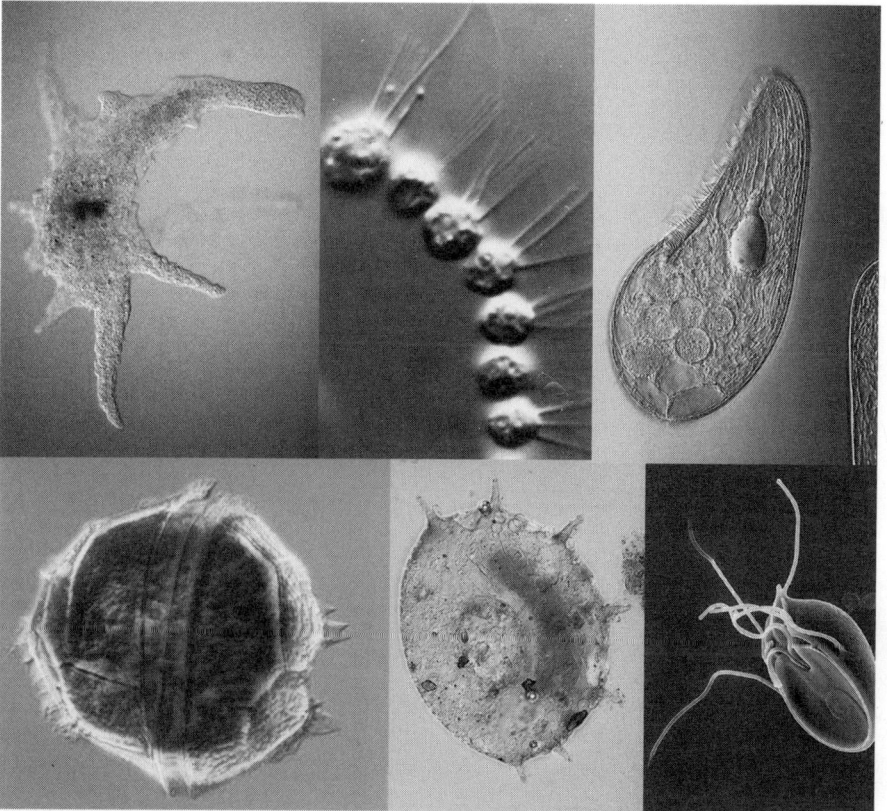

A variety of different protozoa. Credit: Frank Fox, Sergey Karpov, Dr. Stan Erlandsen, Thierry Arnet, Tsukii Yuuji, https://commons.wikimedia.org/wiki/File: Protozoa_collage_2.jpg

called pseudopods. Their main body then flows into the pseudopod, propelling them forward. This process is repeated over and over again as they move from one place to the other. The ability to change shape allows them to explore very tiny spaces as they look for food. This group is more predominant in clay soil or soil with tiny pores.

Testate amoebae are recognized by their hard shell called a "test." They are more abundant in sand and silty soils where the soil pores are larger. These guys also move by creating a pseudopod. As it grows, it pulls the rest of itself, including the test, along the path of the pseudopod, mimicking a moving snail.

The Ciliate Group

Cilia are tiny hairs that can be moved like oars and they give this group their name. The cilia surround their body and allow them to move in any direction. These hairs are also important for moving bacteria-laden water past their bodies. They are facultative organisms that can live in low oxygen levels and are often an indicator of anaerobic conditions. They are more populous in wet conditions.

The Flagellate Group

These are the most abundant protozoa and they move by agitating their flagella, a whip-like structure. They are very small, with a very flexible body, so they can access pore spaces that larger protozoa can't reach. Flagellates are true aerobes, and their presence in soil indicates a good high oxygen level (above 6 ppm). They are more prominent in dry soil.

Protozoa Causing Human Diseases

There are some well-known human diseases caused by protozoa. Malaria is a protozoan infection of the red blood cells that is transmitted by the bite of a female mosquito. Amoebiasis and giardiasis are both intestinal illnesses caused by infection from stool or an infected water system. African Trypanosomiasis, also known as "sleeping sickness," is caused by microscopic parasites of the species *Trypanosoma brucei*, which is transmitted by the tsetse fly.

Life Cycle of Protozoa

Protozoa have two life stages. In normal conditions, they exist in an active stage known as the trophozoite stage. In this stage they grow, feed, and reproduce. If conditions become inhospitable, they enter into a cyst stage by forming a thick, protective wall that allows them to hibernate and resist the environment until it improves.

Protozoa can reproduce sexually or asexually, but most free-living protozoa in soil reproduce asexually by binary fission. Once the organism gets large enough, it starts duplicating all of its organs, including its DNA. The cell wall is then constricted in the center until it divides the cell into two daughter organisms. They can reproduce every eight hours.

The life cycle of parasitic protozoa can be much more complicated and may involve more than one specific host.

Nutrient Cycling

Protozoa need much less nitrogen than bacteria, and it builds up inside of them. So, what do they do? They excrete it into the soil solution in the form of ammonium. Ammonium can be used by plants directly, but much of it is converted to nitrate by bacteria and then used by plants. This production of plant available nitrogen is critical for plant growth.

Protozoa-Plant Interactions

In addition to nutrient cycling, protozoa have a direct impact on root growth.

Free-living rhizobium bacteria in the rhizosphere provide roots with a source of nitrogen. It has been known for some time that their presence has a direct impact on root architecture. What is only now becoming clear is that microfaunal grazing by protozoa may be an even more important factor.

Roots get thicker and branch more with a high population of protozoa. Grazing keeps the number of bacteria lower, which makes it easier for roots to compete with microbes for available nitrogen, resulting in increased root and shoot growth in plants.

Amoebae also produce plant hormones that stimulate root branching and root elongation. Why should they do this? More roots translate into a higher root surface area and more exudates. The extra exudates result in better growth of bacteria, which, as you know, is the main food source for protozoa. Protozoa encourages plants to grow more roots so they have more to eat.

Protozoa can even affect pests on above-ground plant structures. Barley grown in soil with high levels of protozoa grew stronger plants and produced higher yields but also had a higher number of aphids. The healthier plants seemed to tolerate and ignore the higher pest level.

Plant Diseases

Protozoa are mostly beneficial to plants, but there are cases where they cause diseases.

The flagellate *Phytomonas leptovasorum* causes phloem necrosis in coffee plants. This protozoan lives in the sieve tubes of the phloem and eventually kills the plant. A similar disease infecting the sieve tubes of palm trees is caused by *Phytomonas staheli*. Protozoa are transmitted from tree to tree by a beetle.

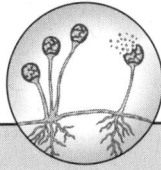

CHAPTER 8

Viruses

VIRUSES ARE INFECTIOUS PATHOGENS that are extremely small even compared to bacteria, and are not made up of cells. The simplest virus is just a small piece of genetic material surrounded by a protein coat. Even the genetic material is minimal and usually only codes for a couple of protein genes. In comparison, a plant's DNA codes for forty-five thousand genes.

Are viruses alive? That is a debated question. Most scientists do not consider them living, but they are not dead either. Viruses don't have the normal organs of a cell and don't maintain their internal environment. They don't grow and they can't reproduce on their own. They lack protein-synthesizing and energy-producing apparatus that are considered a requirement for a living organism.

Viruses are able to use a host cell to carry out many of the functions living organisms perform, such as replication. They also use the energy from their host cell. They are able to adapt to their environment, and in fact they can mutate very quickly, as we have all experienced with COVID-19. I have called them "infectious pathogens," which is an appropriate term for most of them, but new science indicates that some may be beneficial to plants.

Viruses have a range of effects on plants. In some cases, the plant is hardly affected at all while at the other extreme, the plant dies quickly. Infected crops may show reduced growth and lower yields.

Viruses are obligate parasites, which means that they need the cellular machinery of a host to function. They are able to survive for some time outside of their host but only as an inactive entity.

Most are fairly short-lived, but some, such as TMV (Tobacco mosaic virus), can survive for decades.

Most life forms including animals, plants, fungi, and bacteria are hosts to viruses, but each virus species usually infects only one type of host. They are responsible for many important plant diseases and cause significant losses to crop yields.

Viruses are immobile outside of their host and rely on other organisms or the gardener to move them around.

In their mature form, they are referred to as virus particles or virions. Plant viruses have either a helical rod shape or a spherical shape, with diameters of about 30 nanometers (0.00003 mm). For comparison, a typical leaf mesophyll cell is 1,500 times larger.

The genetic material in most organisms consists of double-stranded DNA (double helix). Some plant viruses use a single strand of DNA but most use RNA. RNA is simpler to replicate and has the advantage of being more resistant to UV light.

Typical infected leaf symptoms include mosaic patterns, chlorotic or necrotic lesions, yellowing, stripes or streaks, and leaf rolling or curling. Flowers can be deformed in shape or color. Dramatic color breaking (streaking or blotching) is very common. Fruit symptoms may include mosaic patterns, stunting, discoloration, or malformation. Chlorotic ringspots can also form. Stems can develop pitting, grooving, and tumors. Many virus symptoms look similar to those caused by abiotic issues, and correct diagnosis usually requires laboratory tests.

Viruses in Soil

Viruses have been studied in animals, humans, plants, and even the ocean, but very little is known about them in soil. Recent work shows that soil can contain many kinds of RNA viruses that likely infect fungi and bacteria, but they can also infect plants and animals. The viral populations change quickly in response to environmental changes.

Viruses have important implications for agriculture, and scientists believe viruses to be major drivers of biogeochemical processes (turnover of one chemical form to another) in soil.

Unfortunately, little is known about how gardening activities affect the viral population but because of their sheer numbers, the effect is probably significant since any changes to the viral population will trickle up through other microbes and then to plants and animals.

Most plant viruses are transmitted by insect vectors that invade the host plants through the aerial parts, but there is also considerable infection through the roots from soil-inhabiting vectors such as plasmodiophorids, chytrids (related to fungi), and nematodes. Many cause serious diseases in crop plants, including the beet necrotic yellow vein virus (BNYVV) that infects sugar beets, the potato mop-top virus (PMTV), which causes brown rings in tubers, and the wheat mosaic virus that infects cereal crops.

Life Cycle of a Virus

Virions outside of a host don't do very much. They are immobile, they don't eat, and don't metabolise very much since they lack the molecules and cellular equipment to do what most cells do. Life literally starts for them once they enter their host cell. For plant cells, this is a passive process where the virion enters the cytoplasm of the host cell through damage in the cuticle and cell wall. Once inside a suitable host, they start replicating.

Since the virus does not have the biological tools to replicate genetic material, it hijacks the host tools and uses them. In effect, it tricks the host into replicating its genetic material. The virus also uses the host to make proteins and other molecules that are needed to make complete, new copies of itself.

Once the host cell is full of new virions it is time to infect other cells and this can be done in one of two ways. Some viruses leave after they kill the host cell, while others leave by using a budding process where they pass through the outer membrane of the host cell without directly killing the host cell.

It takes a couple of hours for a virus to multiply in a cell and move on to the next cell. The virus now needs to find other host cells. Many viruses will produce special "movement proteins" that help it get to a neighboring plant cell.

A virus can also speed up the process of infecting a whole plant by entering the vascular system of the plant. This is normally done through the phloem using the sieve tubes, where virions move along with other photosynthate molecules. The phloem is used as a superhighway for virions to reach all parts of the plant.

At various points around the plant, the virion will leave the phloem and infect a local cell. It can then use cell-to-cell movement to infect all of the cells in that area. The time needed to infect a whole plant after initial contact can take anywhere from a few days to a few weeks.

The Spread of Viruses

Viruses infect just about every species of wild and cultivated plant and there are no cures available to gardeners. The only defense is to prevent its spread. Plant viruses are spread in one of two ways: natural vectors and gardening techniques.

Natural Vectors

Natural vectors can be a wide range of plant-feeding organisms including arthropods, nematodes, plant-parasitic fungi, and insects. Aphids and whiteflies are responsible for transmitting many virus species.

Each type of virus is transmitted by one vector type. So, aphids do not generally spread the same virus as white flies and it is even more specific than that. For example, a specific virus might only be transmitted by a specific aphid species. What this means for the gardener is that aphids on an infected plant may not be spreading the virus because that species of aphid does not live on other types of plants.

Vector transmission can be a very simple process. In some cases, the vector takes a bite out of an infected plant and the virion hitches a ride on the insect's mouthparts. When the vector bites a new plant, the virion enters the new host plant.

The *Potyvirus* genus produces a special protein called "helper component" that glues the virions to aphid stylets (inserting mouthpart). Infection of a new plant takes from a few seconds to a few minutes after the aphid starts feeding.

Transmission can also be a more complicated affair, where the virus also infects the vector. A good example of this is the tomato spotted wilt virus and its thrips vector. The virus is ingested by the thrip as it feeds on the infected plant. It then infects the thrip and travels with it, infecting plants for the rest of its life.

Controlling vectors in the garden can go a long way toward preventing virus infection. A better option is to remove infected plants.

Spread by Gardeners

Gardeners spread viruses in three different ways: pruning, propagation, and failure to remove infected plants.

An important part of preventing the spread of viruses is being able to recognize an infected plant. Unfortunately, that is not

Microbe Myth: Bleach Is Good for Cleaning Pruners

It is a good idea to sterilize pruners between plants so that you don't transmit diseases, especially viruses. People love using household products in the garden, but it's not always a good idea. Bleach looks like a good candidate, but it's not.

Bleach does kill microbes, but it is also corrosive to metal. It causes rusting and pitting of your pruners and loppers, requiring you to sharpen them more often. The pitting on the blade can become infested with microbes, which means the tool has to sit in the bleach longer to be disinfected. Oh, and that causes more pitting.

Bleach can cause serious damage to eyes, it stains clothing, and it disintegrates your gardening gloves. It is also phytotoxic and can damage plant cells around the cut.

Vinegar is also recommended a lot on social media, but it's a poor choice because it does not kill microbes very effectively.

Trisodium phosphate (TSP) does disinfect but like bleach, it will damage the metal blades.

Rubbing alcohol (70 percent isopropyl alcohol) is best and easy to use. Just spray it on the pruners and start cutting.

easy to do. Plants can be infected and show no symptoms at all. Symptoms may not be visible early in the year, only to develop by midsummer. Many symptoms also look like abiotic issues and are misidentified.

Since there is no cure for an infected plant, the best thing to do is to remove the plant and don't compost it. Place it in a plastic bag and put it in the trash bin. It is even a good idea to take a bit of soil from around the roots and discard that with the plant.

Viruses are spread through plant juices, so cleaning pruning tools as you move from plant to plant also works well. Alcohol, chlorine bleach, trisodium phosphate (TSP), and pine oil can all be used to sterilize tools, but the best option is rubbing alcohol containing 70 percent isopropyl alcohol. It can be used right from the bottle and does not require a soaking time.

Propagating Infected Plants

Never propagate a suspect plant. Viruses can be spread by all of the vegetative methods including division, cuttings, and grafting.

Both pollen and seed can be infected with viruses. For example, the cucumber mosaic virus (CMV) is one of the approximately 20 percent of all known plant viruses that can be transmitted by seed. Seed left to ripen on infected plants is easily spread by birds, which leads to more infected plants.

Virus Infection of Microbes

In addition to plants and animals, viruses also infect other microbes.

Fungi

Mycovirus is the term used for viruses that infect fungi, and such infections can be harmful to plants. As we will discuss later, the interaction between fungi and plant roots can be extremely beneficial to plants. If the mycovirus harms the fungi, it will indirectly harm the plant.

However, such infections can also be beneficial to plants. A number of mycoviruses have been identified that infect pathogenic fungi, which is good for plants. In fact, this looks like such promising science that researchers are working on developing mycovirus strains to combat fungal diseases in plants.[12]

Chestnut blight was devastating to chestnut trees in North America and Europe and was caused by a fungus called *Cryphonectria parasitica*. It is now known that a mycovirus, *Cryphonectria hypovirus*, reduces the virulence of the fungus and keeps it from killing trees. This biocontrol agent was naturally introduced into Europe, probably along with the fungi. It has now spread to most tree populations in Europe, and in some cases artificial infection was attempted. It has also now been introduced in a limited way in North America.

Bacteria

Did you know bacteria get sick just like us? It turns out that most of the viruses in the world infect bacteria, not plants or people. These viruses are called bacteriophages, or phages, which literally means "bacteria eaters." There are around 10^{30} different viruses in the ocean (a 1 with thirty zeros) and most of these infect bacteria.

Electron microscopy has provided scientists a means of visualizing phages, some of which appear to have "heads," "legs," and "tails." They may look like walking robots, but they are nonmotile and depend upon Brownian motion to reach their targets.

If you have kept your eye on techniques used to manipulate genetic material, you may have noticed a term called "CRISPR." It is a newer and more efficient method than GMO for altering the DNA of organisms.

What has this got to do with bacteriophages and bacteria? Quite a bit. Bacteria use natural CRISPR to defend themselves against viral infections by using it to chop up the viral genetic material before infection happens. Scientists learned about CRISPR by studying how bacteria defend themselves, and they then adopted the technique for other cell types.

Pathogenic bacteria are one of the major pathogen types to cause diseases in diverse plants, resulting in negative effects on plant growth and crop yield. Chemical bactericides have been used to control them, but the appearance of resistant bacteria as well as their potential negative effects on the environment has led scientists to look for other control agents. Bacteriophages may meet the need without any of the negative side effects.

A number of bacteriophages have been found that control plant-pathogenic bacteria, thereby reducing the disease impact on plants. These phages are very specific to a small number of clones and therefore will not harm the general bacterial population. Application of these to plants can be done as foliar sprays or soil drenches.

Foliar sprays are not very successful since the phages are easily destroyed by high temperature, high and low pH, and sunlight irradiation (UVA and UVB). Soil drenches are more effective but require a high level of water in the soil to keep them active and moving toward their host. Most of these techniques are still experimental.

Virus Infection of Plants

Tulips

One of the earliest records of a plant virus was the tulip break virus (TBV), which was described in 1576. At that time, only the symptoms were known. The virus was not suspected until bulb manipulation experiments were done in 1928, which showed that the condition could be transferred from plant to plant using some tissue from the infected plant. TBV causes color breaks in the flowers, giving rise to really cool-looking tulips.

You might have heard of Tulip Mania, a crazy time in Dutch history where tulips were being auctioned off at extraordinary prices. The story starts around the 1590s when Carolus Clusius, who was an avid bulb grower, established a botanical garden at the University of Leiden and cultivated his precious tulips. He noticed that some had color breaks, blooming with exotic streaks of color, and these rarities became very desirable and commanded high prices.

The price for individual bulbs reached ridiculous prices by the early 1600s, and only the wealthy could afford them. Keep in mind that nobody suspected a virus, but the infection did lead to weak bulbs and fewer offspring. This only made them even rarer and

Tulip "Semper Augustus"

more expensive. Tulip Mania reached its peak between 1633 and 1637. A special tulip, Semper Augustus, had beautiful red streaking and is reported to have been the most expensive bulb ever sold.

As with all speculation, it all came to a sudden crash in 1637. Prices plummeted, speculators lost their shirts, and the world was whole again. Semper Augustus became weaker and weaker due to the virus and is no longer available today. TBV is not extinct and can reinfect bulbs, so it is illegal to have tulips with color breaks in Holland.

Tulip Mania did happen, but much of the story is fiction. It only affected a small number of very wealthy people, not the whole society. Prices were high for a bulb, but the people involved were very rich, so they didn't care, and only a few bulbs were sold at these high prices. The event did not impact the "whole" Dutch economy as so many claim. It is a fun story, though.

Hosta Virus X (HVX)

Hosta virus X is a serious problem that infects many popular hosta cultivars. It was discovered in 1996, and has spread throughout the USA and is now a global problem. Initially, growers found unusual hostas in their garden and thought they had found a new cultivar. Hostas such as "Break Dance," "Eternal Father," "Leopard Frog," "Blue Freckles," and "Lunacy" were not actually new cultivars but instead were hostas infected with HVX. Sounds a lot like Tulip Mania.

The disease is a bit tricky to identify, but unusual blotchy spots or streaks as well as leaf puckering are indicative of virus X. For subtle symptoms, it is useful to hold an infected leaf up to a light and compare it to a healthy leaf. The brighter light makes the symptoms more visible. One of the problems in identifying the disease is that some varieties don't show symptoms, others only show them on older plants and many only show them later in summer.

I have found two such plants and in both cases the symptoms only showed themselves once the newly purchased plant had grown for a couple of months in the garden.

There is no treatment. If you have a suspect plant, get rid of it as described above.

Tomato Viruses

More than twenty viruses infect tomato plants including the following:

- tobacco mosaic virus (TMV)
- tomato mosaic virus (ToMV)
- tomato spotted wilt virus (TSWV)
- pepino mosaic virus (PepMV)
- cucumber mosaic virus (CMV)

Symptoms include lighter green mosaic patterns, brown-streaking and distortions to the leaves, reduced leaf size called "fern leaf," and marbling patterns (uneven ripening) on the fruit. Herbicide damage can also cause some of these symptoms. The only way to have a positive identification is to have the plant tested.

Resistant varieties are available for some tomato viruses, but they won't completely protect the plant from the virus. Look for abbreviations such as ToMV and TMV in your seed catalog.

Rapid Mutation

Everyone is now intimately familiar with the coronavirus which causes COVID-19, and its ability to spread and mutate quickly. Many viruses do the same thing, and it's the reason we have so many different viruses on Earth.

How does a virus mutate?

It really is no different than any other mutation. The virus infects a host cell and uses their machinery to replicate its genetic material. This process is not a perfect system, and from time to time a mistake is made. That mistake is a mutation and as it is replicated further, it is used to create new virus particles. These then infect other hosts and get further replicated. Provided that the new virion is viable, it spreads far and wide.

The fact that viruses replicate very quickly and in large numbers results in lots of mutations, leading to a large amount of genetic diversity both within and between species.

Beneficial Interaction with Plants

Most of our virus knowledge is about parasitic ones, but some viruses are beneficial to plants, especially in extreme environments. They can improve a plant's tolerance to drought or extremes of cold and hot soil temperatures.

The Tobacco mosaic virus makes several crops more tolerant of drought and freezing temperatures. The white clover cryptic virus can suppress nodulation in legumes when adequate nitrogen is present, thereby allowing the plant to use resources more efficiently.

Viruses that produce mild symptoms have been used to protect plants from more deadly ones. There are also viruses found in plants that seem to be harmless and we do not yet know their beneficial effects. These are found in both wild plants and crop plants including peppers, rice, beans, carrots, figs, and radish.

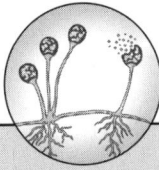

CHAPTER 9

More Microbes

THERE IS A LOT OF DISCUSSION about bacteria and fungi, but there are many other types of organisms in soil. The following are some of the more important ones.

Archaea

Archaea (pronounced like Ikea with an "a" instead of an "i") are important single-celled organisms that are similar to bacteria and some people lump them in with bacteria, but science separates the two groups.

Their shape and size are similar to bacteria, but some of their genes and metabolic pathways are more similar to eukaryotes (animals, plants, fungi). Archaea are able to use a wider range of energy sources than eukaryotes, which includes things such as sugar, ammonia, metal ions, and even hydrogen gas. The salt-tolerant *Haloarchaea* uses sunlight as an energy source, but it does not fix carbon like plants.

Some archaea are classified as extremophiles, which live in environments such as hot springs and salt lakes, and a special type dubbed ARMAN lives in acid mine drainage. They have been found in every type of environment, and along with plankton, they may be one of the most abundant organisms on the planet. Up to 10 percent of microbial cells in temperate soils are likely archaea and they associate with plants both above and below ground.

In the human microbiome, they are found in the gut, mouth, and on skin. In soil, they play an important role in carbon fixing

77

(converting inorganic carbon to organic carbon), nitrogen cycling, decomposition of organic matter, and maintaining microbial symbiotic communities. There are no known pathogenic archaea.

If these guys are so common and important, why have you not heard of them before? They are difficult to detect, and science is only now finding them because genetic analysis is more prevalent.

Actinomycetes

Actinomycetes are called filamentous bacteria, or mold bacteria. They look and grow like fungi but are biologically similar to bacteria. They grow hyphae-like threads that consume resistant organic matter, and they are tolerant of dry soil, alkaline soil, and high-temperature conditions. They produce antibiotics such as streptomycin and actinomycin that stop the growth of other microbes. Some of these are available as commercial drugs.

Actinomycetes tend to be found in decaying organic matter. They can protect plant roots from disease, and in a few cases they cause diseases such as potato scab. Their affinity for higher temperatures and their ability to decompose tough organic matter makes them an important contributor in hot composting. These organisms are responsible for the earthy smell of damp, well-aerated soil.

Frankia are special actinomycetes that form symbiotic nitrogen-fixing nodules on over two hundred species of nonleguminous plants. Most are trees and shrubs such as alders, sea buckthorn, and *Casuarina* species, many of which are pioneer species that grow in very poor soil.

Why Does Soil Smell So Good?

The earthy smell that rises after a summer rain is so pleasant that we use it to scent perfumes. One of the compounds responsible for this fragrance is geosmin, and its biology in soil is very interesting.

Geosmin is made mostly by an actinomycetes called *Streptomyces*, which makes a host of other compounds including the well known antibiotic streptomycin. Scientists have been wondering for quite

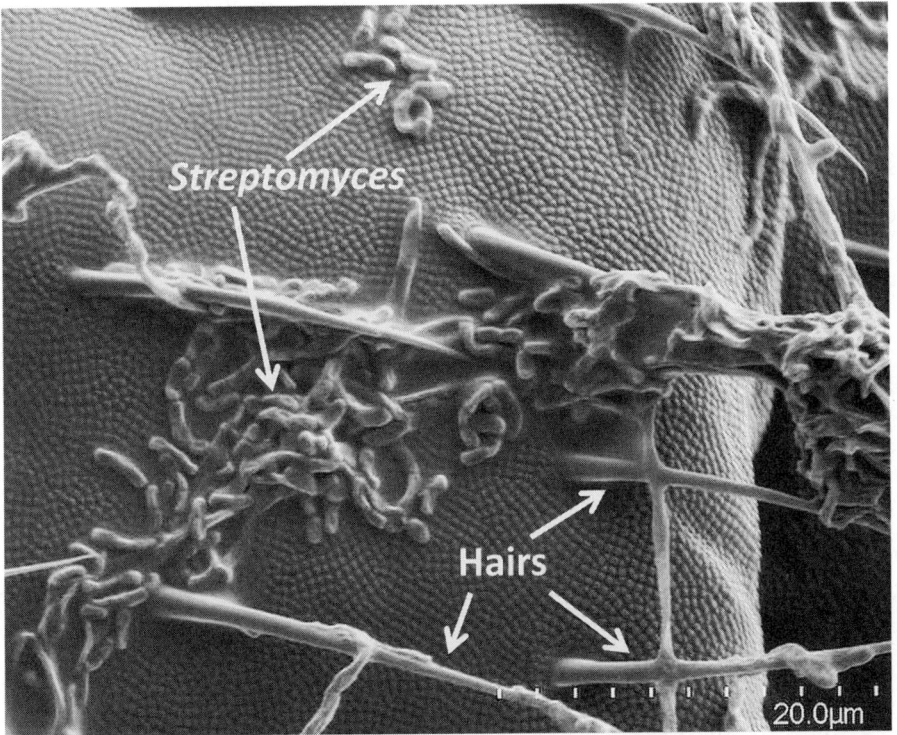

Streptomyces spores hitching a ride on the surface of a springtail. Credit: Ola Gustafsson, Lund University

some time why this organism makes geosmin, and they now have some insight into this.

Streptomyces grow mycelium threads through the pores in soil. When they run out of nutrients, it is time to move on, and they form spores. These spores spread by wind and water. As part of the process of producing spores they also produce geosmin, which attracts springtails.

Springtails are tiny critters (hexapods) that get their name from their ability to hop around like fleas, and they have a particular attraction for geosmin. As the springtails rummage through soil looking for geosmin, the *Streptomyces* spores get attached to the springtails who then spread them throughout the soil. Springtails are jumping Ubers for actinomycetes.

Microbe Myth:
Soil Is an Antidepressant That Makes You Feel Good

You probably saw the error-riddled meme that says, "Soil is an antidepressant. The smell of *mycobacterium vacii,* a microorganism found in soil, compost, and leaf mold, lights up neurotransmitters that release serotonin, a mood-lifting hormone." [13]

Gardeners were quick to accept these facts, but I had a close look at the science behind this. First of all, there is no such organism; however, *Mycobacterium vaccae* does exist.

Serotonin has been well studied and is known as the "feel-good hormone." It plays a key role in staving off anxiety and depression. So, if the smell from microbes in soil causes higher levels of serotonin, it is quite likely that these smells make us feel good. However, the pleasant smell of soil is due to geosmin which is made by *Streptomyces* bacteria, not *Mycobacterium vaccae*. It is also not linked to serotonin levels.

There is no scientific evidence that exposure to *Mycobacterium vaccae* or geosmin changes our serotonin levels. As with most gardening memes, there is a smattering of truth, but most of the message is wrong. Don't believe gardening memes.

Cyanobacteria

Cyanobacteria are commonly called blue-green algae, although they are not scientifically classified as algae. The name cyanobacteria refers to their bluish color, although this can range from green to blue depending on the species. They use a variety of photosynthetic pigments such as carotenoids and chlorophyll to fix carbon from CO_2.

They appear to have originated in fresh water and are considered to be the first organism to produce oxygen as a by-product of photosynthesis. Photosynthesis in plants takes place inside an organ called a plastid (e.g., chloroplast) and it is believed these are the remnants of ancient cyanobacteria that were integrated into plants millions of years ago.

Cyanobacteria produce some of the most powerful natural poisons known as cyanotoxins. These bacteria can cause algae blooms, which then cause shellfish poisoning due to these toxins.

Some blue-green algae are able to fix nitrogen, but their contribution to soil nitrogen is small.

Algae

Soil algae are single-celled organisms that photosynthesize like plants and can be green, blue-green, or brown in color. Some live on their own in soil and others live in combination with fungi to form lichens.

A gram of soil can contain ten thousand to one hundred thousand algal organisms, and because of their light requirements, most are found on or near the surface of the soil. They also need a fair amount of moisture to grow well. Being photosynthetic, they produce 20 to 30 percent of the oxygen we breathe and are found

Microbe Myth: Diatomaceous Earth Repels Slugs

Diatomaceous earth (DE) is available in two different grades: a food grade and a pool filter grade. For horticultural purposes, you should always use the food grade or a product marketed for pests. The pool grade is manufactured differently, and will not work to repel slugs or insects.

In horticulture, diatomaceous earth is used as a pesticide. When it is applied to insects, it removes their waxy protective coating, which leads to dehydration and death. Some people think it scratches their hard coating, but that is unlikely the cause of death.

It does repel slugs provided it is dry, but once it gets wet they just crawl all over it. It is claimed that the sharp edges of the diatoms cut the foot (i.e., the bottom) of the slug and then the slug either bleeds to death or dehydrates due to a loss of moisture, but that is all a myth. DE might have sharp edges, but they are so small it does not cut the slug, who happens to have a very tough foot that is used to crawling all over pointy things.

mostly in fresh and salt water. Little is known about terrestrial species.

Diatoms are a special type of algae that have an outer wall made of transparent silica. This skeleton forms thousands of different intricate patterns that help identify the various species. It is estimated that there are between twenty thousand and two million species, and scientists are discovering new species every year.

Gardeners have been exposed to diatoms through a product called diatomaceous earth (DE). DE is made from a soft silica-based rock with particle sizes in the range of ten to two hundred micrometers, or a fraction of a millimeter. The rock consists of fossilized diatoms.

You might be surprised that kelp is also an alga and it can grow up to 160 feet (50 m) in length.

Very little is known about the relationship between algae and plants, but new studies point to a close relationship where algae provide significant benefits to plants, and they have even been found inside plant cells.

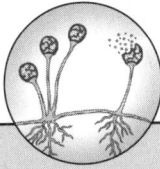

CHAPTER 10

Microbe Communities

EACH OF THE PREVIOUS CHAPTERS deals with a single type of microbe, and that is a good way to learn about these organisms, but it doesn't paint a realistic picture. In the real world, these organisms do not live isolated from each other. Instead, they form real-world communities and each community contains a mixture of different species and types. More importantly, the various types of microbes have very close interactions with each other. Plants are affected by the sum effects of the community, and not by a single organism.

Microbes are difficult enough to study. Can you imagine the complexity of trying to understand a whole community, especially when we have not even identified the majority of species in that community? It is one of the reasons that scientists are just now looking at this. DNA analysis has helped a lot. It allows scientists to look at a community and describe it in terms of different genes, rather than actual species.

Plants also play a vital role. Microbes affect the plants, and the plants affect the microbe community. You really can't separate one from the other but to keep things simple, I will deal with the interactions between microbes in this chapter, and then look at microbe-plant interactions in the next three chapters.

Microbe Communities and Microbiomes

Picture a large, busy city at midday. The sidewalks are full of people rushing by each other. Cars are zooming down the street. A truck comes by every now and then. Bicycles weave in and out of traffic. Some people are at home having lunch while others are sleeping after a long midnight shift. Now imagine that all of these entities are microscopic microbes. Some are large while others are quite small. Some move quickly, and others not at all. Each microbe tries to eke out a living, but they are constantly affected by all of the other microbes around them. That is a microbe community.

When this community exists in a specific location such as a single leaf, or beside the roots of an apple tree, it is known as a microbiome. You may have heard of the gut microbiome, which is located inside your digestive tract and it includes all of the microbes in the gut.

Each area of a plant has a different microbiome. The organisms on leaves will be different from the ones on fruit or on roots. Individual plants of the same species will have similar microbiomes,

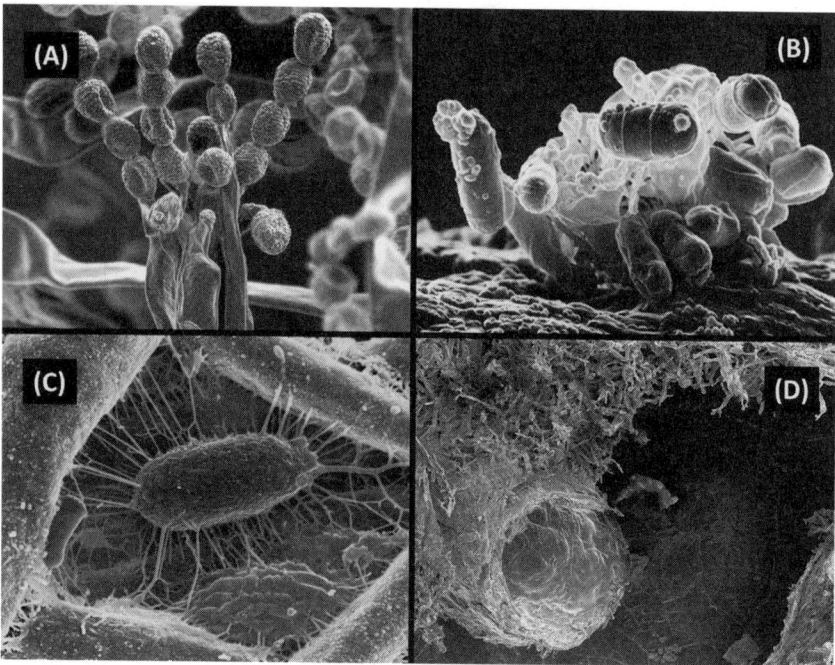

(A) spores (conidia) of *Penicillium* spp. in rhizosphere, (B) cluster of *E. coli* bacteria, (C) microbe attached to plant root, (D) pine tree root surrounded by a soil fungus. Credit: (A) EMSL, (B) Microbe World, (C) & (D) Pacific Northwest National Laboratory[14]

but they won't be identical, in the same way that each of us has a different gut microbiome.

You might think that neighboring plant roots would have the same microbes because they grow in the same soil, but roots from different species produce different exudates, which in turn cultivate different microbes. We tend to think of microbes as benign organisms that are just hanging around, occasionally causing problems with our plants, but nothing could be further from the truth. They are constantly attacking each other, and the spoils of these fights are generally beneficial to plants.

The relationship between microbes is much more complex than first thought. As scientists focus more on communities as opposed to individuals, they are finding that many organisms benefit from symbiotic relationships with others. The presence of one species of microbe may have a large impact on the population of a second species, and this can be both positive or negative.

Signaling Between Microbes

How does a microbe know who is sitting beside it? How do they know if their neighbor is a friend or a predator, and how do microbes find a sexual partner?

Microbes use a system of chemical signals. I know some people will anthropomorphize (attribute human qualities) this and call it "communication," but it's not communication in the human sense. Communication requires the sending and receiving party to have knowledge of each other and to understand the meaning of the message. Microbes don't have either of these characteristics.

Signaling between microbes is a simple matter of creating various chemical compounds and excreting them into the liquid that surrounds themselves. In some cases, these compounds are excreted indiscriminately. In other cases, they are created and excreted after the organism comes in contact with a triggering molecule. The trigger compounds interact with the cell's outer membrane and cause an involuntary chemical reaction.

This is similar to your reaction to an unpleasant odor. You suddenly sense the odor, and you might turn up your nose or you might move away from it. You may not know where the scent is coming from or who or what produced it, but you react to it. This is signaling, and not communication.

These signals can cause an organism to move. If the signal indicates food, they might move toward it. If it smells like a predator, they move away from it. In either case, the microbe's reaction is involuntary—it does not really understand the difference between food and a predator.

These signals can have an even greater effect when they activate the genetic material of the organism. Most of the DNA and RNA genes' code for proteins and chemical signals can turn on the cellular processes to start making a very specific protein. This process is not unique to microbes. All living organisms do this inside their cells. In most cases, the signaling compounds are chemicals produced inside their own bodies, but even in humans, external chemical signals can cause the production of specific proteins.

What do these proteins do? Many proteins are enzymes, and enzymes cause chemical reactions to happen. Almost nothing happens inside an organism without the help of an enzyme. Consider this example. The scent of a predator is received by the organism and it activates the production of an enzyme that makes a toxic compound. This toxin is secreted into the water around the organism, repelling the predator. The toxin makes the microbe poisonous to its predators.

Alternatively, the signaling compound may originate from food, some dead organic matter, or even a living microbe. The microbe then produces digestive enzymes and secretes them toward the potential food.

In the case of sexual reproduction, different sexes of the organism produce different pheromones to indicate their presence to the opposite sex. This is the same method used by insects and higher-order animals.

How does a microbe know who is sitting beside it? It doesn't. These are simple organisms that have no brain or the ability to know things. All of their actions and interactions with each other can be explained by chemical signals.

Microbes Attack Microbes

Many microbes are meat eaters and prey on other microbes. The larger ones swallow the smaller ones whole, but even very small microbes can eat each other by secreting digestive enzymes through their outer membrane. They spray these juices on the enemy, and once digested they slurp up the liquid food.

Nutrients are the spoils of fighting microbes and they are always in limited supply, especially carbon and nitrogen. Microbes use chemical warfare to try to hoard these resources by secreting chemical repellents and toxins to keep others away.

Killer Fungi

Since fungi grow and function similar to plants you might think that they won't hunt prey, and that is true of most of them but not all. Some fungi are able to attack living prey. About 150 different species are able to capture nematodes, either with sticky adhesive pads similar to a sundew plant, or a noose-like loop that snares and captures them. Once captured, the fungi exude enzymes to digest their living prey.

The oyster mushroom is a common delicacy, but it is also a vicious hunter. They grow on trees and dead fallen wood. The wood provides them with a high carbon source, but it lacks the nitrogen needed to fully digest it. Nematodes make a good nitrogen meal.

The oyster mushroom sets a trap by exuding chemicals that smell like dinner to a nematode. When the nematode gets close enough, it bumps into specialized fungal filaments that are tipped with a toxin that paralyzes them. The fungus then grows hyphae into the nematode and digests it from the inside out.

The fungi *Drechslerella anchonia* sets lasso-like snares to catch nematodes. Once the nematode enters the snare, the fungi tightens the grip so it can't get away, and it slowly digests its prey.

Fungi are not without their own predators. They are attacked by a variety of organisms, and their main line of defence is chemical warfare. They exude toxins, insecticides, and antibiotics that keep predators away, prevent them from developing correctly, and even kill them. After all, dead predators make a good meal.

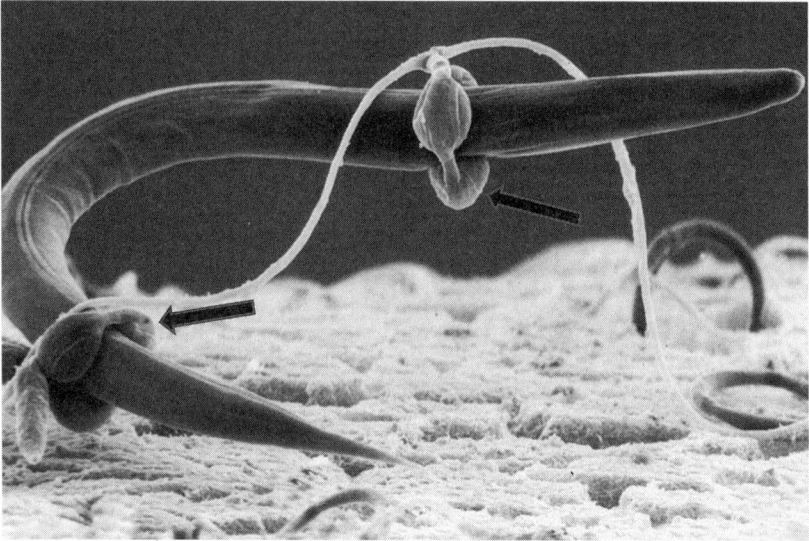

The fungi *Drechslerella anchonia* uses lassoes to catch nematodes.
Credit: Nancy Allin, George Barron, University of Guelph

Their fruiting bodies can be quite toxic and if you eat the wrong mushroom, the fungi can even kill you. Dogs are particularly susceptible to mushroom toxins as are rodents, but each animal reacts differently to a specific toxin. For example, deer mushrooms cause gastric distress in most people, but rodents love them.

Killer Yeast

Are you afraid of attacking yeast? You should be if you are a yeast or other fungi. A special clone of our common brewer's yeast (*Saccharomyces cerevisiae*) is able to secrete a toxic protein that is lethal to sensitive cells of the same species and to other fungi. The yeast that secretes the toxin is immune to it.

These killers were first discovered in 1963 as scientists investigated spoiled beer. What we now know is that it is not the natural yeast DNA that is causing the problem, but some RNA from a virus that codes for the toxic protein. The virus causes the yeast cell to unknowingly produce the toxin, which then harms susceptible members of its own species.

Since the protein that makes the toxin also affects other types of fungi, it has been used to treat some fungal diseases.

Bacteria, Protozoa, and Nematodes

The interaction between bacteria, protozoa, and nematodes is one of the most critical relationships in soil and is responsible for many of the nutrients that are available to plants.

Bacteria are very plentiful in soil; they are small enough to be eaten easily and I assume they are pretty tasty, at least to protozoa and nematodes. They are also high in nitrogen, making them especially valuable in the food chain.

Bacteria don't move very far or fast, so their main line of defense is chemical warfare. They produce toxic substances to keep predators away. Their small size also lets them hide in small openings in the soil that are too small for some of their predators.

Protozoa feed by selecting individual bacteria, while nematodes tend to just suck in large amounts of bacteria at the same time, similar to the way a whale eats small shrimp. They are both aquatic organisms and live in the soil solution, but their difference in size, motility, resting stages, and reproductive strategies mean that the environment influences them differently.

For example, protozoa are much more important in heavy clay soil than nematodes because they can move around in smaller pore spaces. In this environment, their selective grazing favors plant-inhabiting bacteria to the detriment of plants. In other soil types nematodes are more prevalent, which can be more beneficial to plants.

Not all bacteria are equal prey. Some are easier to catch by protozoa, and others by nematodes. This creates an interesting relationship between these three organisms. The species of each type has a profound effect on the total community that exists in any particular area. Bacteria that are more mobile have a better chance of evading the slower-moving protozoa, but they can't outrun the nematodes. Nematodes prefer open spaces due to both their size and the way in which they feed. Free-floating bacteria are a better prey for them than bacteria that stick tightly to soil.

Large bacteria are the same size as smaller protozoa, so protozoa are not a threat to them. The toxin produced by some bacteria will harm protozoa, and so they ignore these types, but nematodes are not affected as much by the small amount of toxin a single bacterium can produce. There is also the added complexity that nematodes also eat protozoa. In fact, when they engulf a large amount of "small bits of food" many protozoa are included in that meal.

What does all this mean? The soil environment, including the type of soil, its physical structure, the amount of aggregation, and even the amount of retained moisture all have a significant influence on the amount and type of protozoa and nematodes living in any one spot. This in turn has a significant effect on the number and species of bacteria living in the same spot.

These microbe communities in turn have a big impact on the plant species that grow. From a gardening perspective, it is almost always better to have more of all three types of organisms.

Mycorrhizal Interactions

Mycorrhizal fungi form symbiotic connections with plants, and they may even connect plants together through the common mycorrhizal network. They are an important part of any microbiome because they also interact with all of the other players.

Bacteria-Mycorrhizal Interactions

Mycorrhizal fungi depend very much on the exudates produced by plant roots—it is their main carbon source. In some cases the fungi competes with bacteria for this rich resource, but in other cases the fungi stimulates bacterial activity by sharing its food, and in turn the bacteria help the fungus.

Fungi produce a chemical called trehalose, which attracts a group of bacteria known as the Mycorrhiza Helper Bacteria (MHB). One of these MHBs, called *Pseudomonas fluorescens* excretes thiamine (vitamin B) which helps the fungi grow. The presence of this bacteria in eucalyptus seedlings produced a 300 percent increase in plant growth.

It has been suggested that MHBs induce the plant to produce growth hormones, which indirectly improves the growth of the fungi. Some MHB are very specific to a few fungal species and others are more generalists, but they don't seem to be plant specific. They enhance mycorrhizal function, growth, nutrient uptake, and help repel some pathogens. These bacteria are naturally present in soil where they form complex interactions with fungi as plant root development takes place.

Bacteria have also been found inside fungal hyphae and even inside fungal spores. It is believed that the bacteria help spores germinate.

Fungal-Mycorrhizal Interactions

There is a lot of talk about mycorrhizal fungi, because they have been studied the most, but it is important to remember that there are many other types of fungi and many are important to the gardener.

Pathogenic fungi are ones that cause diseases in plants and generally attach and feed off them. Saprophytic fungi decompose nonliving organic matter by producing enzymes that decompose cellulose, hemicellulose, and pectin, material most organisms can't touch. Both of these fungi are present in the rhizosphere, but they have to compete for space with mycorrhizal fungi. Ectomycorrhiza (EM) are also able to degrade most organic matter, but they are much less efficient than the saprophytic fungi. They also produce organic acids that inhibit saprophytic fungi, allowing EM to dominate. This increase in EM results in slower decomposition of organic matter and lower amounts of nutrients.

Mycorrhizal fungi also interact with pathogenic fungi and help plants grow better when an attack does occur. The mechanism for this is not well understood, but it includes improved plant nutrition, changes in the chemical composition of plant tissues, and changes in the rhizosphere bacterial community.

Arbuscular mycorrhiza (AM) reduces the incidence of soil-borne fungal diseases. It is not clear if this is due to an interaction between the AM fungi and the pathogen or if it is the result of increased nutrient availability to the plant. Both could be at play. AM does not cause a specific defense response, but it does speed up the response in plants. It is important that colonization happens before the pathogen attacks the plant.

Animal-Mycorrhizal Interactions

Springtails (*Collembola*) are small animals that are normally less than a quarter inch (6 mm) in size and live in soil and leaf litter. They can't digest the organic matter, but they are important in the process because they chop up larger organic matter into smaller pieces, which are then more efficiently decomposed by smaller microbes.

Collembola also feeds on fungus. These hyphal grazers seem to prefer saprotrophic and parasitic fungi over mycorrhizal fungi, but they will feed on the latter. This type of grazing does not seem to impact plant growth, although this has not been studied extensively.

But fungi don't just sit around waiting to be eaten. The EM fungus *Laccaria bicolor* actively kills and consumes springtails and then passes on the excess nitrogen to their hosts, such as the eastern white pine (*Pinus strobus*). When these seedlings are colonized by the fungus, as much as 25 percent of the nitrogen in the seedlings comes from digested springtails.

AM fungi keeps the population of some nematode species low while other species seem to grow better in the presence of AM fungi.

Earthworms ingest all kinds of organic matter, and in the process they ingest fungal propagules which are then spread throughout soil in their castings.

This book talks mostly about microbes and their interactions with each other, but it is important to understand that very similar relationships exist between microbes and other small life. These animals then interact with even larger ones and so on up the food chain. Even quite large animals, such as deer and skunks, rummage in soil or eat plants and thereby pick up microbe hitchhikers that are moved to new locations.

Lichens

Lichens look similar to fungi, but they are actually a symbiotic association between three organisms: a fungus, a green alga or a cyanobacterium, and yeast.

The association between the fungi and the algae/bacteria has been known for 150 years, but it has only recently been determined that yeast are an important part of the organism. This example demonstrates how difficult it is to discover some of these tiny organisms.

The discovery of the yeast started with an investigation into why two lichen species seemed genetically identical but looked very different. One was yellow and produced a toxic acid called vulpinic acid, while the second was dark brown and produced no acid. Both were made up of the same fungus and alga. An in-depth analysis of the genetic material revealed the presence of another actor—a relatively unknown group of yeasts.

The yeast was found embedded in the lichen cortex, or "skin." Some additional internal bacteria have also now been found.

Taxonomists are still debating the best way to describe this organism, but maybe we should not think of it as a single organism. Lichen are more properly described as a community of organisms rather than a simple algae-fungi association.

In most cases, the fungal partner is unable to grow without its symbionts, and the relationship is mutually beneficial. The algae or cyanobacteria is able to photosynthesize and make sugars that

Variety of different lichen on a branch.

are excreted into the community. Fungi absorb water and nutrients, and provide much-needed shade for the light-sensitive algae growing under it. The yeast helps lichens ward off predators and repel microbes.

Lichens are long-lived and grow relatively slowly, and there are still unknowns about how they propagate but most can do it vegetatively. The organism is routinely found on woody plants and some people will scrape them off, believing that they are parasitic and harm the tree, but this does more damage than leaving them alone. Lichens do not harm the plant they are growing on. They are actually good to have in the garden because they are very sensitive to pollution. Their presence indicates good air quality.

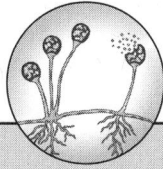

CHAPTER 11

Plants Love Microbes

MOST MICROBES DO NOT LIVE on or near plants but there is certainly a higher concentration near them and for good reason. Plants actively farm the microbes next to them. They provide a physical place for them to live and they modify the area to make the space more habitable.

What is surprising to many gardeners is that plants actually cultivate microbes inside themselves. The internal parts of plants are home to many microbes. The space protects them and feeds them. The plant makes use of the chemical by-products from the microbes. Both plants and their symbionts help each other keep pathogens away. It is a win-win for both. Plant microbe interactions are critical for healthy plants.

Nutrient Availability

One of the key benefits of microbes growing near plants is that they make nutrients more available to plants. Fungi have an easier time extracting phosphorus from soil than plants, so plants use fungi for this purpose. Bacteria are better able to dissolve rocks and release nutrients for plants. The activity of microbes around roots lowers the pH, which helps release nutrients from soil.

Nitrogen and phosphorus are the two most critical nutrients for plant growth because nitrogen is almost always in short supply and phosphorus is difficult to extract from soil. I'll focus the discussion on these two nutrients, but many of the concepts also apply to other nutrients.

Nitrogen

Nitrogen is critical to the growth of plants and microbes, and it is usually the nutrient that is rate limiting. All organisms are trying to get as much of it as they can, and they all compete with each other.

It is ironic that nitrogen gas makes up 78 percent of our air and yet that nitrogen is useless to most organisms. Nitrogen gas has to be converted to compounds such as urea, ammonium, nitrite, and nitrate before most organisms can use it. The nitrogen cycle shows how various organisms convert one form of nitrogen to another, and this is fully explained in my book *Soil Science for Gardeners*. Here's a summary of that process.

Nitrogen-fixing is the process of converting nitrogen gas to usable forms of nitrogen, and the only organisms that can do this are a select number of bacteria. These were discussed in detail in chapter three on bacteria. Nitrogen-fixing bacteria produce ammonium. Other bacteria then take over and convert ammonium to nitrite and then to nitrate. Bacteria and fungi prefer using ammonium, while plants prefer nitrate.

Our demand for more food has resulted in the use of more fertilizer because natural systems can't provide enough nitrogen for our food needs. Unfortunately, the addition of nitrogen fertilizer inhibits natural nitrogen fixation. The negative consequence of this is that natural systems shut down and become even more ineffective, which in turn means we need to use even more fertilizer.

There is also another source of nitrogen and that is mineralization by microbes, the process of degrading organic matter to release nitrogen. The extent of this process depends on the amount of available organic matter, the concentration of microbes, and environmental conditions. For example, microbe activity is very dependent on temperature. Mineralization is five times higher at 77°F (25°C) than 40°F (5°C).

Nitrogen mineralization is one of the main benefits of adding compost to gardens each year. Even finished compost takes about five years for the material to be completely degraded, and during that time there is a steady release of nitrogen. As microbes die, they are added to the pool of organic matter. Plants shed leaves and roots and also contribute to this process. The added compost ignites the process and helps keep it going, but it is really the growth of microbes and plants that form the main fuel for future mineralization.

Phosphorus

Phosphorus (P) is also a macronutrient, and it can be rate limiting for plant growth. It turns out that most soil contains lots of phosphorus, but much of it is tied up in forms that plants and even microbes have trouble using. The amount of available P depends on soil characteristics and microbial activity, but it is in the order of 2 percent of the total amount.

Phosphorus compounds tend to be insoluble, which means they easily precipitate out of water as solids. Think of the coating inside your kettle. Once they are solids, most organisms can't use them but some special microbes can.

To compensate for low available P, agriculture adds more fertilizer containing phosphorus, but this does not work very well. Only about 15 percent of the applied phosphorus is used by plants. The rest either gets converted to insoluble forms that plants can't use or washes into waterways, causing significant ecological harm.

Phosphate Solubilizing Microbes

Phosphate solubilizing microbes (PSM) are very efficient at producing available phosphorus by mineralizing organic P, solubilizing inorganic P minerals, and storing P in their biomass. PSM can be bacteria, fungi, or actinomycetes, but bacterial candidates have been studied the most.

Phosphorus is an important part of many biological molecules, including ATP and DNA. Living microbes contain a fair amount of these compounds in their tissues. When they die or are consumed, the phosphorus is released into the phosphorus cycle as organic P. PSM are able to convert this organic P into soluble P for plants.

Both PSM and nitrogen-fixing bacteria improve plant growth by making nutrients available, but when the two are both present, they supercharge this growth. When the plant is no longer starved for phosphorus, the nitrogen-fixing bacteria become more active and produce higher levels of nitrogen.

Most of this research work has been done in the lab, but scientists are now trying to show these effects in crops. It is hoped that applying the right cocktail of bacteria and fungi might improve both phosphorus and nitrogen levels and result in lower fertilizer requirements.

The Phyllosphere

The phyllosphere is the total above-ground surface of the plant and its associated microbes. This is clearly an important part of a plant but, from a knowledge perspective, we know much less about its microbes than the root system. A Google Scholar search for "phyllosphere plant health" produced twenty thousand results while a search for "rhizosphere plant health" produced one hundred seventy thousand results.

The above-ground parts of a healthy plant are colonized by a variety of bacteria, yeasts, fungi, algae, and, less frequently, protozoa and nematodes. Bacteria are by far the most numerous colonists of leaves, although many of the diseases affecting leaves are fungal. Bacteria populations on leaves can be as high as sixty-five million per square inch (ten million per square centimeter), or almost a billion cells per gram.

The microbial communities are diverse and include many different genera. They vary between plant species and also change throughout the growing season. The physicochemical environment on the surface of a leaf is quite different from the environment around roots, and therefore the microbes found on leaves are also quite different. For example, pigmented bacteria, which are rarely found in the rhizosphere, dominate above-ground plant surfaces probably because they can handle the intense solar radiation better. Common root bacteria such as *Rhizobium* spp. are rarely found on plants above ground.

Strictly speaking, the phyllosphere includes only the outer surface of the plant, but I will include the internal part of the leaf in this section because the microbes on the surface are closely related to what happens inside the leaf.

Surface Leaf Microbiome

The leaf surface looks flat and smooth to us, but at a microscopic level it is quite irregular. To a small microbe, it is mountainous with many valleys that offer protection. These valleys also trap water for longer periods of time. The stomata are relatively large openings, and they expel moisture that tends to remain trapped in these valleys, at least for a little while.

The surface of a plant is a hostile environment for any microbe. It is normally hot during the day and cold at night. Rain washes small cells and spores off the surface, and moisture dries up quickly. The sun beams down high-intensity UV light that fries the little guys. The conditions are not the most extreme on the planet, but they do change quickly. Even if a microbe survives all of this, they still have to contend with a limited food supply.

How do microbes survive these conditions?

Microbes use a survival tactic similar to that of large grazing animals such as the caribou and antelope, who live in herds. They live in large aggregates, which offer protection to the ones on the inside of the clump. When things get really dry, the outer ones die and form a kind of vapor barrier for the ones living inside the clump. Predators, heat, and cold affect the outer ones the most.

A leaf may seem dry to us, but it has a very thin coating of water on its surface. From a microbe's point of view, things are not as dry as we think. Drought does make life on the leaf harder, and some species start dying off while others actually increase in numbers. Population dynamics are constantly fluctuating as the environment changes. The high-stress, nutrient-poor condition of the leaf surface is an important selection mechanism.

UV light damages DNA, and if enough gets damaged it will kill the cell. Incidentally, this is what I researched for my master's degree. The microbes that live in the phyllosphere tend to be ones that have a high capacity for repairing their damaged DNA.

Protection from UV light can also be provided by making special pigments that protect against the light, a kind of sunscreen for microbes. Others contain a pigment called rhodopsin that functions like chlorophyll and allows them to fix carbon from CO_2 for a food source.

Most microbes still have a problem obtaining enough food, but the plant helps out with this. Plant leaves leak small amounts of sugars, providing microbes with an energy source. These sugars are not evenly distributed, and some areas are better for microbes than others.

Another food source is the arrival of new microbes. Remember that many microbes are quite small and are easily moved from place to place by wind or raindrops. Many of these, such as fungal spores, are in a dormant state. When they land on a leaf they become food for the established microbes. Other microbes land on the leaf but can't survive the harsh conditions. They soon die and feed others.

Free-living nitrogen-fixing bacteria do live on leaves, but little is known about their importance for providing nitrogen to either the plant or other microbes. It is known that in natural settings these organisms contribute nitrogen to Douglas fir needles.

Leaf surface bacteria can produce surfactants, which form a biofilm to help reduce moisture loss. Some are more motile, allowing them to essentially hide under other microbes for protection.

How do microbes get on the plant in the first place?

It all starts with seed germination, as I'll discuss later in this chapter. The seed contributes the initial microbes. As the first sprout pushes its way through soil to get some light, it also picks up more microbes from the soil. Once above ground, spores and microbes land on the leaf as they fall from taller plants, are washed off of other plants by rain, and many can travel long distances on wind currents. The number and diversity increases as the seedling grows. As the cool spring turns to hot summer, the environment plays a big role in determining population changes.

If natural bacteria are sprayed onto leaves, they can become the dominant species. This is very evident in hydroponics where the

Microbe Myth:
Keep Plant Leaves Dry to Prevent Powdery Mildew

Powdery mildew is a fungal infection that produces a white coating on leaves. The most commonly suggested method for preventing this disease is to keep plants dry. Thin out branches on shrubs so you get better air flow and faster evaporation. Don't spray leaves when you are watering.

Science tells us the truth about suggestions like this.

Powdery mildew is spread by spores floating through the air. The spore produces a feeding tube and inserts it into the leaf. The fungus then extracts food through these feeding tubes, similar to a mosquito sucking blood from your arm. This food allows the fungus to grow outside the leaf and form the white mycelium we see. When the fungus reaches maturity, it forms new spores.

What condition causes the spread of the disease? High humidity favors spore formation, while low humidity favors spore dispersal. It is also important to remember that this disease is actually caused by many different species, and they are all a little different.

The reality is that wet leaves dry quite quickly. Try it for yourself. Spray some leaves with water and time how long it takes for the water to evaporate. Thinning out shrubs is a good idea but has limited effect on humidity. There is very little a gardener can do to control humidity, which is mostly a function of the weather.

The best thing you can do to eliminate powdery mildew is to grow resistant plants.

natural level of microbes on leaves is very low. Spraying the right microbes in such conditions results in a higher level of microbes, which then help plants grow.

The individual microbes on leaves are not visible to the naked eye but, once they form larger colonies they can be seen in some cases. One case that is familiar to gardeners is powdery mildew, which is described in the sidebar.

So far the discussion is about microbes living on the leaf, but we now know they also live inside the leaf, the internal leaf microbiome.

Internal Leaf Microbiome

To our macro eyes, the inside of a leaf looks fairly solid. It is fleshy, but you don't see a lot of air gaps. On a microscopic level, the inside of the leaf is quite hollow with lots of hiding spaces for microbes.

Leaves also have pores in their outer epidermis called stomata. For a microbe, the opening in a stomata is an appealing place since there is a regular flow of moisture and oxygen out of the leaf. It also provides easy access to the apoplast (air space inside the leaf).

It may seem as if the stomata make the plant very vulnerable to microbes, but that is not the case. Plant leaves have something called immunity receptors that can sense the presence of bacterial pathogens. This causes the plant to produce specific proteins that initiate a defense attack on the pathogen while at the same time triggering the closure of the stomata, preventing more pathogens from entering the leaf.

Closing the stomata does cause a problem for the plant. It prevents the expulsion of excess oxygen and the absorption of more carbon dioxide. Without CO_2, the plant has to shut down photosynthesis until it is again safe to open the stomata. Plants do have many stomata on a leaf, and they can be shut down individually.

It is clear that plants control which microbes live inside the leaf, and scientists have identified specific plant genes that manage this. When these genes are removed by creating special mutations, the plant loses control of the microbes. Such leaves have high levels of internal microbes and the population is very different from normal leaves. The changes cause dramatic visible changes as well, including areas of dead cells and general yellowing. A gardener who sees these leaves would normally blame a disease, but it may just be a case of the leaf losing control over its microbiome.

The microbes that live inside a leaf are normally beneficial or neutral, but pathogens can also enter and live inside the leaf. Bacteria and viruses can cause significant damage.

Floral Microbiome

Microbial communities on flowers can have an important impact on plants since it allows them to gain entry into the fruit. Once in fruit they can affect our food, and some will enter the seed to be transmitted to future generations.

Bacteria and fungi are present early in a flower's life, well before the petals open. They exist both on the outside and inside of the buds.

As flowers open, the stigma and hypanthia (the floral cup containing the nectar) contain a number of compounds including sugars, proteins, and fats that are attractive to microbes. The number of microbes increases over time due to both newcomers and growth of those that arrived early. For example, the monkey flower (*Mimulus aurantiacus*) has yeast growing in 20 percent of day-old flowers and 70 percent of older flowers. The hummingbirds that pollinate this flower help spread the yeast.

Microbes can also be found on petals and pollen. There are some bacteria that are only found on flowers and nowhere else on the plant.

Animal visits to flowers significantly increase the number of microbes. In some cases, flowers that have not been visited by pollinators have no yeast, while visited flowers have a high level in their nectar. Bee feces containing microbes are frequently detected on flowers.

Bees and other pollinators are not that different from us in that we all have microbes growing on us and in our gut. If a bee gets a drink of nectar they are sure to leave some "germs" behind because they don't follow the no-double-dipping rule. They also pick up microbes from the flower as they brush by it, and it has been shown that some of the natural bacteria found on flowers are beneficial to bees.

Honeybees carry a number of viruses, and these are left on flowers after a visit. It has now been shown that these viruses can be subsequently picked up by native bees who then become infected. Gardeners have learned to love honeybees, but they are actually harming our native populations. By the way, the cultured honeybees are not native to most of the world.

The fragrances produced by flowers are volatile compounds that can be beneficial for some microbes while also inhibiting the growth of others. Flowers also produce a number of other metabolites that don't have a fragrance, and we are just starting to learn about their effect on the floral microbiome. I wonder how perfume and deodorant scents affect microbes? Deodorants used on armpits do reduce the microbiome there, but it is not clear how much of that is due to scent.

Some floral bacteria use the fragrance compounds as food. It is now known that the bacteria also produce a scent of their own, which mixes with the natural scent of the flower. A flower grown in sterile conditions smells different than one grown in the garden due to the odors produced by microbes.

What is the benefit of bacterial scents to flowers? To test this, researchers added these natural floral bacteria to field-grown rapeseed plants and found that treated plants had more visits from pollinators and produced more seed. It seems that the bacterial fragrances enhance the quality of the flower fragrances, at least for pollinators, and benefit the plant with higher seed production.

Most of the microbes on flowers do not seem to benefit the plant.

Gardeners Affect the Phyllosphere

I hope that by now two significant points are becoming clear:

1. All parts of a plant are covered in microbes, some of which are harmful, some are neutral, and many are beneficial.
2. A gardener can't see the microbes or identify the beneficial ones.

Less obvious is the fact that everything you do in the garden affects the microbiomes. Watering, feeding, cultivating soil, and even touching the plants affects your microbes. Most of these activities won't do a lot of harm, but spraying to fight pests and diseases will. There are so many home remedies out there and most people think that anything from the kitchen is safe. It's not.

Microbe Myth: Product X Only Kills the Bad Bugs

This is a common belief of gardeners. Be it insects or microbes, they believe that a product can only kill the "bad" bugs. This is almost never true.

The whole idea of bad bugs and good bugs is a human construct. We think some are good and others bad based on what we perceive to be the right way to live. We like bees but hate wasps. We love monarch butterfly larvae and hate larvae eating our roses.

Nature has a different perspective. All bugs are good bugs, and they are all important members of the whole community. Every one of them adds some value, even if we don't know what that value is.

The biology of a given type of organism is very similar. Bacteria are all very similar. Good bacteria and bad bacteria react the same to most treatments. Alcohol kills almost all of them. Bleach does the same. Spraying plants with a soap solution will kill both good and bad bugs.

There are a few treatments that only affect a certain species of bug. For example, Japanese beetle traps only attract Japanese beetles. But for the most part, sprays and chemicals affect both good and bad bugs.

The other thing to be aware of is that most garden sprays are used to kill something we can see, but they also kill the microbes we can't see.

Things from the kitchen are still chemicals and each chemical does potential harm. In the case of kitchen products they almost all harm microbes. Oils, baking soda, dish soap, hand soap, and even Epsom salts will harm them. They are even disrupted by just spraying water on the leaves.

What should a gardener do when they have a problem? Step one is to do nothing. Take some time to assess the situation. Do you really need to solve the problem? Many inexperienced gardeners feel they need to take immediate action, but that is rarely true. More experienced gardeners know better and just let nature take care of things.

I grow some three thousand different plants and I almost never spray anything. If I get aphids, I wait a bit and ladybugs show up. If

I get powdery mildew on plants, I ignore it—it never kills the plant. I also get it on cucumbers, and last year I tried a mildew-resistant variety. Home gardeners should rarely have to spray their plants with either home DIY solutions or commercial pesticides.

If you do spray, understand that you are harming the phyllosphere microbiome—you just don't know how serious that harm is. Too bad we can't hear the tiny squeals of microbes as they die!

Microbiome on Commercial Crops

Is there even such a thing as a healthy plant microbiome in today's agricultural fields, with acres of identical plants assaulted by pesticides, herbicides, and fertilizer?

Agriculture affects plant microbes in a variety of ways. Sprayed chemicals will certainly alter the microbiome on any crop. Some microbes will die and others will thrive. In most cases, we know almost nothing about these changes.

Breeding new varieties of crops also affects the microbes. Breeding may change the leaf surfaces; it definitely changes the internal biochemistry, which in turn affects the chemicals plants give off to both attract microbes and deter them.

Large monocultures also affect them. In nature, microbes associate with more than one type of plant, but with monocultures there is only one option for them. There is also the issue of microbes finding the next generation of plants. If they don't travel well and are not associated with seeds, it is unlikely that the right microbes even exist in the new field.

None of this sounds very positive for agricultural crops, but it may not be as bad as you think. A team at Berkeley, University of California, identified and domesticated (grew them in a lab) the main microbes found on commercial tomatoes. These microbes were stable and formed a healthy microbiome that was effective at fending off random microbes that landed on the plant.

The microbes on commercial plants may be different from the ones on their native cousins, but they are effective. That is good news because it means that we should be able to manipulate the microbiome with inoculants to make healthier crops that need less fertilizer and less frequent spraying for pests.

Food-Borne Illnesses

The CDC warns that "raw fruits and vegetables contain harmful germs that can make you and your family sick, such as *Salmonella*, *E. coli*, and *Listeria*." You have undoubtedly heard warnings on the news about buying lettuce or tomatoes because of bacteria contamination that causes food-borne illnesses.

The Canadian Government website on food safety says that "fresh fruits and vegetables do not naturally contain microorganisms that cause food poisoning. However, fresh produce can become contaminated in the field through contact with soil, contaminated water, wild or domestic animals, or improperly composted manure. It can also come into contact with harmful microorganisms during and after harvest if it is not properly handled, stored, and transported."[15]

This information is commonly accepted, but it is not entirely true. It is true that various contact points can contaminate food, but the idea that this food does "not naturally contain microorganisms that cause food poisoning" is wrong. It is well documented that several of these pathogens grow on or in the plants long before harvest.

The FDA found the pathogens *Salmonella* spp. and *Shigella* spp. on 4 percent of crops tested. *Salmonella enterica* and *Escherichia coli* have the ability to colonize corn, bean, and cilantro plants under humid conditions but fail to grow in dry conditions. The CDC reports that *E. coli*, norovirus, *Salmonella*, *Listeria*, and *Cyclospora* are found on leafy greens.

It is not known how significant this presence is. Do they cause sickness given that the numbers are usually small? Or is it our harvesting and processing practices that help them grow into numbers that are significant enough to make us sick? We don't know.

Gardeners should assume that all produce from the garden and the supermarket contain these pathogens. Clean the produce well enough so that they don't make you sick. For more information

Microbe Myth:
Soap Is a Good Way to Wash Produce

Fruits and vegetables are washed for several reasons: remove dirt, get rid of pesticides, and eliminate germs.

Removing dirt is a good idea. It can contain pathogens and it doesn't taste very good.

Washing does a poor job of removing pesticides. First of all, the amount of pesticides on the outside of the produce is so small it won't harm you. More than 99.9 percent of the pesticides are natural ones produced by the plant and they are on the inside of produce—washing won't remove them.

Germs such as *Salmonella, E. coli*, and *Listeria* are a real concern. A tap water rinse removes germs as well as or better than a rinse using commercial products, soap solutions, bleach, baking soda, or vinegar solutions. A ten-minute soak in a 2.5 percent solution of vinegar reduces germs by 82 percent and is better than soaking in just water. This is a 1:1 mixture of water to household vinegar.

The Food and Drug Administration (FDA) and Centers for Disease Control (CDC) strongly urge consumers to stick with plain water.[16]

about the best way to wash produce see this article https://www.gardenmyths.com/wash-fruits-vegetables/

Rhizosphere

The rhizosphere is a very thin layer of water, air, and soil right next to the root. It is only 2 to 3 mm thick, but it is crucial to plant growth. It is in this area where all of the concepts in this book come together to create a very unexpected microbiome.

Things start with plant roots. Root growth happens mostly at the very tip of the root, which is covered with a root cap. The root cap is pushed forward through the soil as the root grows longer, resulting in up to ten thousand cells being scraped off each day, only to be replaced by new ones.

A little bit back from the tip is where all of the root hairs grow. These are responsible for picking up water and nutrients, but they don't live very long. After two or three weeks they are old, and the plant sheds them.

The dead root cap cells and root hairs produce a mini-compost pile of dead organic matter right next to the root. What happens to fresh organic matter in soil? Microbes start growing to take advantage of the excess food, and that is exactly what happens in the rhizosphere. The microbiome here explodes in number thanks in part to the growth of plant roots.

The story does not end there. The tip of the root also excretes special chemicals called plant exudates into the rhizosphere. These exudates select for the right kind of microbes and repel pathogens. They also contain a lot of sugar, which adds more carbon to the compost pile, accelerating microbe growth.

The microbe population in the rhizosphere can be a thousand times higher than in the rest of the soil. Microbe predation and death from old age results in even more organic matter being released, and much of this is mineralized quickly into plant available nutrients.

How does all this affect the plant? The roots are now growing in a compost-rich environment. Microbes are processing the organic matter and breaking it down into nutrients that are located right next to the root hairs, making it easy for the plant to absorb them.

The rhizosphere is more than just a good source of nutrients. It is also a protective layer around the roots. The exudates produced by the plant encourage beneficial bacteria to live there and repel pathogens. The bacteria also help out by attacking any pathogens that happen to get near the roots. By cultivating the right bacteria, the plant has created a mostly disease-free environment around the roots.

The most difficult nutrient to get is phosphorus, and plants have a strategy for that too. Why create a lot of long roots when you can get microbes to do that for you? The exudates include signaling compounds that let mycorrhizal fungi know that the plant is ready to get married. The fungi sense the signal and start growing toward the root. When they reach them, they burrow right into the root and form a tight bond with it.

The mycorrhizal fungi consist of long mycelium networks that reach far into the soil to pick up phosphorus and water, which is brought back to the roots. In exchange, the plant supplies the high-carbon sugars needed by the fungi.

It is important to understand that it is the plant that initiates this symbiotic relationship and it is the plant that allows the fungi to enter the roots to make the connection. Plants growing in high-phosphorus soil don't need the fungi and don't make the connection with them.

Which microbes live in the rhizosphere? Just about every microbe mentioned in this book lives there. Bacteria and fungi are key players, but their predators, protozoa and nematodes are also attracted by the easy pickings. Even yeast find a home there.

The rhizosphere is an extremely active environment that benefits both plants and microbes. All of this activity and competition has a dramatic effect on the microbiome, which is very different from the bulk soil.

Plant Root Exudates

Roots release all kinds of chemicals, and as a group they are referred to as root exudates. They are produced for various reasons:

- restrict the growth of competing roots
- attract microbes in order to form symbiotic relationships
- change the chemical and physical properties of the soil and soil solution
- make nutrients more available

The exudates include sugars, carbohydrates, and proteins. Sugars and carbohydrates provide the carbon, and proteins provide the nitrogen—the perfect buffet for living organisms.

The high level of food leads to huge increases in bacteria, which in turn attract their predators. Nematodes and protozoa populations can be twenty-seven and thirty-five times as high, respectively, compared to bulk soil.

Plant roots also produce amino acids, vitamins, organic acids, nucleotides, flavonoids, enzymes, glycosides, auxins, and saponins. These can cause both positive and negative responses from microbes and other plants. Some of these attract specific organisms that plants want to have nearby, while other chemicals can be toxic to certain species.

Amino acids are used to make proteins, and all organisms have a high demand for them. Most organisms, including plants, are able to both absorb and excrete them. In some cases, plants secrete more than they absorb, and this may be of benefit to the surrounding microbes. In other cases, the microbes may be supplying amino acids to plants.

Root exudates are a complex mixture of compounds that vary by plant species, age of the plant, root architecture, environment, time of the year and, the microbes that are present in the rhizosphere. A reduction of leaf surfaces due to pests and diseases affects photosynthesis, and will change the exudates produced.

An interesting example is the effect of temperature. As spring changes to summer and temperatures reach 98°F (37°C), the amount of exudates increases for the warm-growing kidney bean but decreases for the cool-growing pea. The peas are done for the year, so there is no point in producing more exudates. In wheat, seedlings and mature plants can have six times fewer exudates than young growing plants in the tillering stage.

Pesticides and antibiotics sprayed onto plants also affect the exudates, but we know very little about these effects. Excess fertilizer generally reduces the release of exudates because the plants no longer need help from microbes.

Root cells are able to sense certain signaling compounds and respond by secreting appropriate exudates. The actual mechanism of all this is still being elucidated.

One type of exudate that you may have heard about are the allelochemicals that cause negative responses in neighboring plants. The black walnut is the poster child for this phenomenon. It is claimed to excrete juglone, which keeps other plants from growing under it. The reality is that the walnut does not produce juglone, and most plants can grow just fine under a walnut tree. For more on this see *Walnuts, Juglone and Allelopathy* (https://www.garden myths.com/walnuts-juglone-allelopathy/). Allelopathy is real and does happen but not to the extent reported in gardening circles.

Producing exudates is a major drain on plant resources. It is estimated that some plants excrete 50 percent of the carbon they fix through photosynthesis. Annuals excrete 40 percent of their photosynthates, and for the average plant it is around 30 percent.

Producing these exudates is very costly to the plant and so they must get significant benefits from it, and they do. The large number of microbes results in a lot of dead microbes, due both to short life spans and predation. This produces high levels of nutrient ions right next to the roots. This is much more efficient than trying to find and extract nutrients out of soil.

The Rhizosphere pH

Metallic nutrients react differently in soil at different pH. If the pH is too alkaline (above 7.5), iron, boron, manganese, copper, and zinc become unavailable to plants and yet many plants grow just fine in these alkaline conditions. The secret may lie in the rhizosphere. Roots release organic acids and hydrogen ions into alkaline soil. This changes the pH of the rhizosphere and makes nutrients more available.

The large population of organisms in the rhizosphere absorbs oxygen and respires carbon dioxide just like animals. Even roots excrete CO_2 as they carry out respiration. The normal CO_2 level in soil ranges from 0.3 to 5 percent, but in the rhizosphere it can be as high as 20 percent. As this CO_2 dissolves in water, it produces carbonic acid, lowering the pH.

The net effect of all of this is that the rhizosphere is usually more acidic than the rest of the soil, by as much as 2 pH units. This explains why plants can grow in alkaline soil and still get the nutrients they need. The roots are actually growing in acidic conditions even though the bulk soil is alkaline.

The fixation gardeners have on soil pH may be unfounded, at least to some extent. Plants may be able to control their pH environment much better than we think. There are exceptions. Some so-called acid-loving plants don't seem to be able to adjust their pH enough to grow in alkaline soil.

Soil Enzymes

Enzymes are special protein molecules that are responsible for most chemical reactions in a cell. Making large carbohydrates out of simple sugars is done by enzymes. Breaking large proteins into smaller amino acids is also done by them.

The soil solution is full of enzymes. Both microbes and plants excrete them and dead cells break apart, releasing even more. This soup of enzymes can attack microbes, digest organic matter, decompose other enzymes, be absorbed by microbes, and be absorbed by clay particles. An enzyme called urease has been found in permanently frozen soil that is thousands of years old. This concentration of enzymes speeds up the degradation of all the organic material in the rhizosphere. Both plants and microbes benefit.

Mycorrhizosphere (MS)

Mycorrhizal fungi have relatively long hyphae. One end of the hyphae is attached to a root in the rhizosphere, but the majority of the organism resides outside of this region. This larger region is called the mycorrhizosphere.

AM (arbuscular mycorrhiza) accounts for 25 percent of soil microbial biomass and up to 80 percent of the fungal biomass. A gram of soil can contain one hundred feet (30 m) of hyphae. Since most crops form associations with AM the mycorrhizosphere is very important to them.

So far I have described the association between mycorrhizal fungi and plants as being controlled mostly by plants, but newer research has shown that fungi have more control than we first thought.

The fungi are able to direct plants to change the type and amount of exudates plants produce, and this affects the rhizosphere microbiome. For example, when legumes are inoculated with both mycorrhizal fungi and rhizobia bacteria, the degree of nodulation increases and there is a higher amount of fixed nitrogen compared to using rhizobia alone. The presence of fungi also results in an increase of free nitrogen-fixing bacteria and a decrease of denitrifying bacteria.

Exudates are made in the upper level of the plant and moved into the roots, where they have two destinations. They can be excreted into the soil or they can be absorbed directly into the fungal hyphae. AM seem to be able to control how much they get and can draw more than their fair share of exudates out of plants, reducing the amount that goes into the rhizosphere.

Well-developed rhizosheath indicates a healthy soil. Credit: Fred Price, Gothelney Farm, with permission, https://www.gothelneyfarmer.co.uk/

Fungal hyphae are like long tubes, and once the exudates are in the hyphae they move along these tubes to reach all parts of the organism. Fungi use some of the exudates as food and the rest is excreted. This whole process has the effect of moving plant exudates from inside the root to all parts of the mycorrhizosphere.

AM hyphae are also not long-lived and have a turnover rate of five to six days, adding extra organic matter to soil. All of this affects the microbes living next to the hyphae in a similar way to those living in the rhizosphere. The mycorrhizosphere may be much more important to plants, soil, and microbes than we first thought.

You can see the effect of a growing mycorrhizosphere by examining plant roots. The increased growth of hyphae results in better soil aggregation around roots, which is called the rhizosheath.

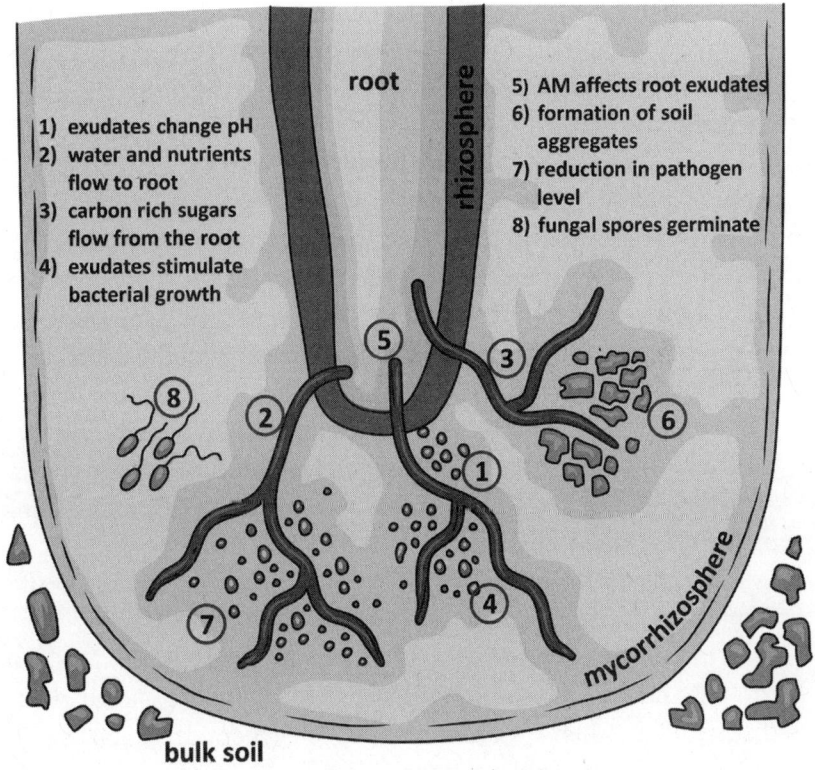

root

rhizosphere

1) exudates change pH
2) water and nutrients flow to root
3) carbon rich sugars flow from the root
4) exudates stimulate bacterial growth

5) AM affects root exudates
6) formation of soil aggregates
7) reduction in pathogen level
8) fungal spores germinate

mycorrhizosphere

bulk soil

Mycorrhizosphere around the root.

If you carefully dig up mature plants that are growing in good soil, you will find that roots look much thicker and browner than they really are.

The amount of soil aggregation in the rhizosheath can be used as a rough indirect measurement of soil health, but results vary due to a number of parameters including the plant species, soil type, moisture in soil, quantity of mycorrhizal fungi, and even the amount of root hairs. Dry soil, more root hairs, and a higher fungal presence results in higher levels of aggregation. Some species such as cultivars in the brassica, allium, and asparagus family do not form rhizosheaths. Rice forms sheaths in drier soil but not in wet conditions.

Rhizophagy Cycle

The rhizophagy cycle is one of the most interesting recent discoveries about plants. We now know that plants attract special microbes and then internalize them so they grow inside the plant. The plant harvests nutrients and other growth-promoting molecules from them and then releases them back into the soil so they can gather more nutrients. In effect plants farm these microbes.

The story starts with the plant producing signally exudates that attract certain bacteria and algae to the surface of their root tips. Once there, the plant internalizes them and they become known as endophytic microbes. The mechanism of this is not yet clear, but at the end of the process the microbes are living inside the roots.

The plant then produces a compound called superoxide that dissolves the outer cell wall of the bacteria, resulting in protoplasts. A protoplast is a fancy name for a cell without a cell wall—think of them as naked microbes. The microbe is intact with all of its internal organs protected inside a cellular membrane, and in this form it can continue an almost normal life.

Without a cell wall the internal compounds easily leak out through the protoplast's membrane. These include basic nutrients that help feed their host as well as special chemicals such as hormones that promote plant growth. The dissolved cell walls contain nitrogen, which also feeds the plant.

It is not all bad news for the microbes since plants want to keep their farm animals healthy. The plant provides sugars and amino acids to help them grow and reproduce. They are also protected from predators, which can't get inside the plant.

As the root tip grows, these microbes find themselves farther back in the root, at the point where root hairs form. As the root hairs form, the protoplasts gather in the tip of the root hair. When they get too crowded, they are expelled from the tip of the root hair to make room for more. Once they are released back into their normal soil environment, they use nutrients and plant exudates to regrow their cell wall. They become invigorated, grow in size, and reproduce.

If they find themselves near another root tip, the cycle starts all over again. This movement of microbes into the root, along the root, and then out of the root hairs is a continual process that is under plant control. We now believe that most, if not all plants, carry out a rhizophagy cycle.

Why have we only discovered this recently? Bacteria are small and colorless, and you can't see them inside a root under a microscope unless you use special dyes that have only recently been used for this purpose. Also, scientists just never looked inside plants for such microbes.

There is one other very interesting aspect to all this. If a seedling is totally sterilized so that there are no microbes in or on the plant, the root does not develop root hairs. If endophytic microbes are added back to the growing media, the plant starts the rhizophagy cycle and makes superoxide. The bacteria try to fight back by producing a plant hormone called nitric oxide, which neutralizes superoxide. This hormone has a side effect: it triggers the elongation of root hairs. The growth of root hairs is triggered by the microbes, not the plant. Plants without endophytic microbes have no root hairs.

I wonder if there are endophytic bacteria for your scalp?

The old concept of plants getting all their nitrogen from nutrients in the soil has to be rethought. In one experiment, as much as 30 percent of their nitrogen came from endophytic bacteria. Endophytic microbes are critical to plant growth, and it has been

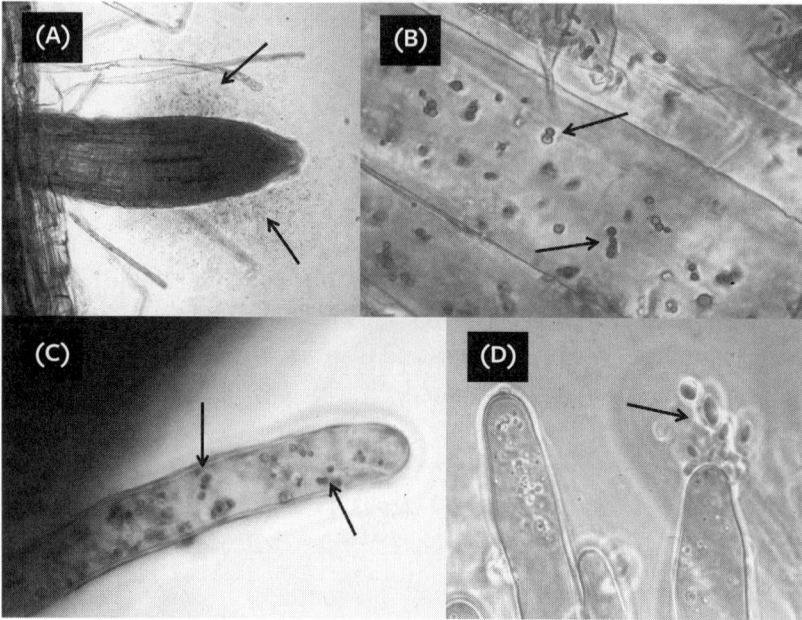

Evidence of the rhizophagy cycle. (A) cloud of bacteria around the tip of a root, (B) protoplasts living and replicating inside the root, (C) bacteria inside a developing root hair, (D) clover root hair expelling endophytic yeast.
Credit: James F. White et al., used with permission[17]

shown that they can be added to agricultural plants to improve crops. For example, adding the right endophytic bacteria will increase the amount of carotene made in carrots, making them a healthier food for us. Other wild types of bacteria that have been lost from modern-day cultivars have been shown to increase yields.

Where does a new seedling get these microbes? Plants are able to get some from the soil, but they also have them on both the outside and inside of the seed. It is one reason for not sterilizing seeds.

Seed Microbiome

How sterile are seeds? Take a whole tomato and sterilize the outside. Remove the seeds in a sterile environment. Are they sterile? Not really. Microbes live inside the fruit, on the seeds, and even

Microbe Myth: Fermentation Should Be Used to Harvest Tomato Seeds

There are three basic ways to collect tomato seeds. The easiest is to squeeze out the seeds, dry them, and store them. A similar way that produces cleaner seeds is to collect them, wash them in water, and then dry them. The third method uses fermentation.

In fermentation, the seeds along with bits of tomato are squished out into a dish and left for several days. The tomato juices ferment and molds grow on the mixture. It is then rinsed through a sieve and the seeds are dried. This produces very clean seeds. Many gardeners believe that fermentation is required in order to make viable seeds.

I tested this a few years ago to see which method worked best, and you can see the results in the reference video.[18] They all produce viable seeds. Fermentation is not required.

How do these cleaning processes affect the natural microbes living on the seeds? Is there a benefit to the plant to leave them on the seeds?

We don't know the answer to these questions, but we should assume that the seed has beneficial microbes on its surface and that it is best to preserve them. Fermentation is an anaerobic process, and it will likely kill any aerobic microbes on the seed. It certainly removes all the bits of dried tomato from the seed, and this might be where the beneficial microbes are hanging out.

Fermentation might result in seedlings missing their natural beneficial microbes.

inside the seeds. The seed microbiome includes epiphytes that live on the surface of the seed and endophytes that colonize inside of the seed.

Now take these seeds and split them into two groups. Do nothing to one group and use it as the control. Take the second group and sterilize the outside of the seeds and then plant both sets of seeds. The seedlings grown from sterilized seeds will have a bigger problem with pathogenic bacteria such as *Pseudomonas* spp. The reason for this is that the natural bacteria on the seed helps it fight off pathogens.

A similar effect is seen when sterile tomato seeds are sprayed with native bacteria. The plants produce higher yields with larger fruit and less disease. This effect is most notable when plants are grown in low-nutrient environments and has almost no effect in highly fertilized crops. The assumption is that nutrient-stressed plants need the help of bacteria and therefore produce more exudates to help them grow, which in turn provides nutrients for the plants. In the presence of more fertilizer the plants no longer need the bacteria, and they stop cultivating them.

The above talks about bacteria, but bacteria, archaea, fungi, and viruses can all be found inside healthy seeds.

The microbe community around a plant is extremely important to its survival, but how does a mother plant pass along the right microbiome to its offspring? In humans, this is much easier. The mother has intimate contact with the baby and through touch and breastfeeding can pass along a good microbe community. Plants don't have this luxury. The only way for them to do this is by allowing the microbes to enter the seed.

How do microbes get into seeds? Some microbes use an internal pathway and travel from the roots or other parts of the mother plant and enter the developing seed along with water and other nutrients. Others arrive externally and hitchhike a ride on pollen or even pollinators to arrive on the stigma before traveling down the pistil. Microbes can also enter seeds after they are dispersed from the mother plant.

The internal volume of most seeds is quite small, has a limited food resource, and is very dry. It is not a great place for microbes to live, and consequently many seeds in a seed pod have no microbes, but some will have a small number of them. The microbe species vary based on plant type as well as the location. For example, seed from Douglas fir growing in one location might have different microbes than those growing in another region. A single Douglas fir seed can have up to twenty-six endophyte species, but it normally has only one or two.

The microbes most commonly seen in seeds come from the mother plant. The mother plays a role in selecting microbes that are beneficial and passes them on to its offspring. These microbes will then grow with the seedling and provide it the same benefits they provided to the mother, namely, easier access to nutrients, relief of abiotic stresses, and protection from pathogens.

Consider this example. The seedlings of cucurbits, including cucumbers, melons, and pumpkins, are prone to infection by soil-borne fungal pathogens. Researchers have now isolated 169 bacterial endophytes from these seeds. These bacteria were tested to see if they prevent growth of various pathogens including *Podosphaera fuliginea,* the cause of cucurbit powdery mildew. Seventy percent showed positive results. Some of these endophytes produce VOCs (acetoin/diacetyl) that are antagonistic to pathogens, and others secreted ribonuclease enzymes that attack pathogens.

The seed of cucurbits come equipped with internal bacteria to fight off fungal pathogens.[19] Pathogens can infect seeds after they are dispersed. You might guess that this would be a big problem for seeds residing in the soil seed bank for months and even years. It does happen, but infection from pathogens is rare not because the seed can't be infected, but because the microbe community in the seed bank contains very few pathogens.

Most of the soil seed bank is near the surface of the soil, and this is also where much of the organic material is found. This organic matter is responsible for a very rich diverse microbial community, and it is able to keep the level of pathogens very low. This seed microbiome system works quite well to control microbes and ensure that only the right ones get passed to future generations, but it's not perfect and pathogens have found ways to use seeds to infect future generations.

Pathogen Migration Via Seeds

Acidovorax citrulli is an interesting pathogen that causes bacterial fruit blotch in cucurbits, especially melon and watermelon. It uses the stigma as an entry point in watermelon and infects the seed. When the seed germinates, the bacteria grows in the cotyledon leaves and then spreads to other parts of the seedling as it grows. Water will wash it off infected leaves and into the soil where it can infect the roots of non-infected seedlings.

If an infected seedling grows into a mature plant, the symptoms show up in leaves and on the fruit. These include water-soaked regions which are followed by necrosis and wilting of stems, leaves, and fruit.

Manipulating the Endophytic Microbiome

Microbes can become part of the seed if they are applied to the flowers. These microbes land on the stigma, travel down the pistil, and enter the developing seed.

Research has shown that this also works if flowers are sprayed manually. This means that with the right concoction of microbes, we should be able to pre-treat seed before it even forms. These living organisms would then travel with the seed and become part of the seedling when it germinates, creating a type of built-in pathogen immunity. This technology could improve seed survival, help them grow faster, and make them more competitive with weeds. It could even make them more drought tolerant, which could become very important as the climate changes.

These organisms would continue helping the plant as it matures and may even become part of the next generation through the seed microbiome.

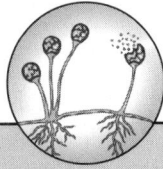

CHAPTER 12

Manipulating Microbes

A LOT OF THE PREVIOUSLY presented information is fascinating, but the real value in this knowledge comes as we learn how to use microbes to grow better plants. That is the focus of a lot of current research.

Fungal-to-Bacterial Ratio

It has been known for some time that the ratio of fungi to bacteria varies in different types of soil and with different plant communities. Forests tend to have a higher ratio of fungi while grassland and agricultural soil tends to be higher in bacteria. This has led to the idea that plants grow best when the fungal-to-bacterial ratio is matched to their needs.

Fungi are relatively large in comparison to bacteria. When we talk about the fungal-to-bacterial (F:B) ratio we are talking about the biomass, not the number of organisms.

In general, the F:B ratio varies depending on plant communities:

- coniferous forest: 100 to 1,000:1
- deciduous forest: 5 to 100:1
- weeds and grasslands: 0.1 to 1:1
- agricultural fields: 0.1 to 1:1

Forests have very healthy soil while agricultural soil is considered to be less healthy. This has led to the idea that healthier soil has a higher fungal population.

These concepts have been further expanded by a number of special interest groups such as Soil Food Web School, Korean

Natural Farming (KNF), and Permaculture who suggest that gardeners should modify their soil F:B ratio to meet the specific needs of the plants they are growing.

Facts About the F:B Ratio

Scientists agree that soil from different natural plant communities have different ratios. It is also accepted that agricultural practices such as tilling and using synthetic fertilizer tend to favor bacterial growth and therefore result in a lower ratio.

The ratios can be modified by changing the way soil is managed. Adding organic matter increases the total microbe population. In general, using organic matter with a high C:N ratio favors fungal growth, and a low C:N ratio favors bacterial growth. Interestingly, adding wood chips (high C:N) either as a mulch or tilled into soil resulted in no change in the F:B ratio. Excessive use of nitrogen reduces the F:B ratio.

The fungal biomass in soil does not change much with a change in pH in the range of four to eight, but bacterial growth is very affected, with bacteria being more prominent at high pH. Conifers prefer an acidic soil, which has a higher F:B ratio than alkaline soil.

When different types of soil from deciduous forests were analyzed, it showed that the F:B ratio varies significantly and is correlated with certain soil properties, which indicates that the ratio is more dependent on soil than on plant type.

What Does Science Say?

The F:B ratio is a snapshot in time and fails to take into account microbial activity. The turnover of bacteria is much faster than that of fungi and it is this turnover activity that is much more important to soil and plants than the relative mass at any given time. For this reason, F:B ratios may not be a useful measurement tool, and researchers are starting to look at other parameters that more closely reflect activity.

I did a search on Google Scholar for scientific studies that looked at plant growth at different F:B ratios. I did not find a single one. I checked for plant lists that showed their "preferred" F:B ratio and found none from the scientific community. A seller of microbes did have a very short list but provided no references to support their claims.

The consensus among scientists with respect to agricultural crops is that the optimum F:B ratio for high yields is unknown. There is a lot of research work trying to understand the relationship between F:B and sequestering of carbon and how F:B influences natural ecosystems. But there is no agreement about the importance of the ratio.

It is clear that we don't know the right ratio for particular plants, nor do we know if it is important. It is quite possible that plants will modify the soil around their roots to suit their needs without any interference from us. Plants drive fungal growth, not the other way around. Any group or company selling products that promote the importance of the F:B ratio for growing plants is completely ignoring current science.

What Came First: The Forest or the Fungi?

We accept the fact that soil in forests has a high fungal-to-bacteria ratio. But does that mean trees need to grow in such a high ratio? What came first, the forest or the fungal-dominated soil? That is a key question that proponents of the F:B ratio ignore.

If we take agricultural soil and modify it to increase the ratio of fungi, will trees grow better? Is the modification required to grow trees? Or is this a case where trees will grow just fine in agricultural soil and over time their presence will modify the soil in favor of fungal growth?

When I first heard the comment "grasses and trees prefer growing in very different F:B ratios," I immediately pictured a modern city park with lots of healthy old trees surrounded by lawns. Both the trees and grass grow well and these parks are located on all kinds of soil. How can these two plant types grow so well together if they require different ratios?

Take a well-used agricultural field and allow nature to take over. What happens? Weeds cover the soil fairly quickly. Soon after that, shrubs and pioneer trees start to grow. The initial trees are deciduous, but eventually the conifers show up. If the trees—and especially the conifers—require high-fungal soils, why would they grow? The agricultural soil and the soil under weeds are bacterial soils and if the F:B ratio claim is true, this soil is unsuitable for trees.

The reason forests have a higher ratio is that trees produce difficult to digest molecules such as lignin. This is not food for bacteria, but it can be used by fungi, who outcompete the bacteria in such an environment, changing the ratio. Grasslands produce much less lignin and much better bacterial food, and therefore bacteria do better, reducing the ratio.

Dr. Ingham of the Soil Food Web School points out that brassicas need a bacterially dominated soil and tomatoes need a more fungal soil. But gardeners have been growing these crops side by side, in the same soil, for years.

The Rodale Institute makes the statement, "weeds require a soil with lots of bacteria." Have they never seen a forest after a fire? The "high fungal" soil is quickly covered in weeds that take advantage of the increased light. Anyone who suggests that weeds can be controlled by changing the ratio has never seen disturbed soil. All soil, no matter the pH or the F:B ratio, is quickly covered in weeds.

Common sense examples don't support the idea that gardeners should change the fungal-to-bacterial ratios to grow plants.

Measuring the F:B Ratio

Given all the talk about the F:B ratio in specialty gardening groups you would think that this is a value that is easily measured, but that is not the case. A meta study that looked at 192 studies found that the measured F:B biomass varied greatly depending on the testing method used.

Microscopy is useful for counting microbes, but it is not a reliable tool for estimating biomass largely because of an overestimation of fungal mass, both due to their size and due to the fact it is difficult to distinguish between dead and living organisms.

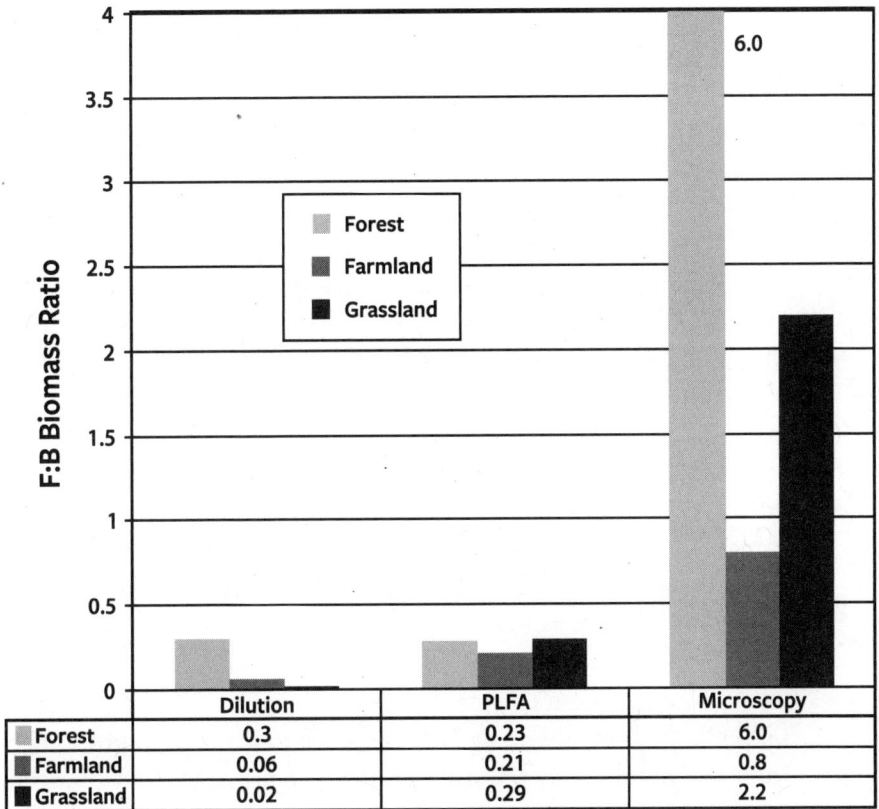

	Dilution	PLFA	Microscopy
Forest	0.3	0.23	6.0
Farmland	0.06	0.21	0.8
Grassland	0.02	0.29	2.2

F:B Ratios measured using three different methods: dilution, PLFA, and microscopy.[20]

There is a new home test kit, called the MicroBIOMETER, that measures the F:B ratio. The test is easy to do and relatively inexpensive ($10 USD). You can learn about this test in this video: *Microbiometer Review* (https://youtu.be/L9azcG2emc0).

Effect of Agriculture

Agriculture has a dramatic effect on soil. It converts very complex plant ecosystems into monocultures, both of which have a dramatic effect on microbiomes. Gardening is a bit less dramatic and can be done in more natural ways, but it still disrupts the natural microbes.

A lot of agriculture grows annual crops, which means the microbes associated with them need to survive short-lived habitats. After harvest, the habitat is lost and microbes have to adapt or perish. This creates a special selection process and limits the available microbes.

What happens to the microbe community as natural land is converted to agriculture?

Imagine a native grassland full of different types of plants. Some grow rapidly in spring, flower, and then die back as others take their place. Summer flowers are replaced by fall-flowering plants. There is a wide range of microbe habitats both above and below ground.

Many of these plants are perennials and have stable rhizosphere habitats that support the same community year after year. Fungi can grow into long filaments without much disturbance. Organic matter slowly builds up year after year, providing a food source for a wide range of microbes.

And then one day the plow shows up. It digs up the soil, destroys fungal filaments, and cuts long established roots into small pieces. A completely new plant is seeded and starts to grow. The local native microbes are still there, but their habitat has changed drastically. Some will adapt and use the crop as their new home. Others will perish. There is generally a decrease in both the total number of microbes and the number of species.

The largest impact of agriculture is on the soil itself. Aggregation is reduced by tilling, organic matter levels change, and chemicals (fertilizers and pesticides) change the chemistry of the soil. Most of these changes happen slowly as virgin land is used for crops, but the changes in microbes are significant, long lasting, and can be measured years after agriculture stops.

Tillage degrades aggregates. This changes the water-holding capacity of soil, which directly affects microbes, who need water to survive. The change in soil pore size makes it more difficult for small microbes to hide from predators.

Increased application of fertilizer causes the nutrient level in the soil solution to increase, making it easier for plants to access nutrients directly. This causes them to reduce the production of exudates and abandon their symbionts, especially fungi. Without the support from plants, mycorrhizal fungi lack a carbon source and die out.

The amounts of microbes in the rhizosphere and phyllosphere drop, making it easier for pathogens to get a foothold. To combat disease, agriculture starts using pesticides, but chemical sprays don't know the difference between a good bug and a bad bug, so the whole microbial community is negatively affected.

Fertilizer use has made our agricultural processes very productive with yields that are higher than ever before. Despite their use, crops still get 40 to 80 percent of their nitrogen from natural soil reservoirs. On average, 50 percent of the applied nitrogen is lost, making the process inefficient and ecologically damaging.

It is important to remember that modern agriculture is able to provide the large amount of food we need to feed everybody. The reason why some areas do not have enough to eat has more to do with distribution than production.

Many people want agriculture to change its ways and return back to a more natural process—a more organic way of farming. That would have some benefits, but it would not produce enough food without converting huge amounts of natural land into farms and that has a significant environmental cost.[21]

A better solution is midway between the extremes. Science needs to find better ways to grow crops while maintaining yields. Efficient probiotics, modified seed that contains a better collection of microbes, microbe mixtures that fight pathogens, and developing new plants that use microbes more efficiently are all good goals.

Effect of Tillage on Fungi

Agricultural soil has a less diverse population that is dominated by a few agricultural-tolerant species. Tillage reduces fungal hyphae and affects spores negatively; and over the long term, some species increase, some decrease, and some are unaffected. For example, wheat in Argentina is colonized by indigenous fungi even after a long history of tillage.

Higher fertilizer levels tend to reduce mycorrhizal fungi as do some fungicides and soil sterilants, but it is more complex than most gardeners think. High phosphorus definitely impacts some arbuscular mycorrhiza species, but not all and moderate levels seem to be beneficial.

Plant breeding also selects for crops that depend less on mycorrhizal fungi and more on synthetic fertilizer because such work is usually done on agricultural soil that contains fewer fungal species.

Fertilizer Affects Microbes

How do fertilizers (both synthetic and organic) affect microbes? As explained in the sidebar, fertilizer does not kill microbes as so many claim, but they do affect populations.

I think the idea that fertilizer kills microbes stems from the fact that agricultural soil generally has fewer microbes than native soil, but there are many factors contributing to this including use of fallow land, overfertilizing, the use of pesticides, inappropriate management of grazing, and loss of plant diversity.

Fertilized plants have an easier time finding nutrients so they waste less energy growing large root systems, and they release fewer exudates. This results in fewer symbiotic relationships with microbes and less microbe food in the rhizosphere. High phosphorus levels reduce the symbiotic associations with fungi, and high nitrogen levels reduce nitrogen-fixing rhizobium nodules.

Both synthetic and organic fertilizers stimulate the growth of some microbes by supplying nutrients, especially nitrogen, which leads to an increase in numbers and improved activity. The increase in biomass can be as high as 15 percent compared to unfertilized agricultural land. Soil bacteria are more sensitive to fertilization practices than fungi, and most species of bacteria are affected positively by fertilization.

Urea and ammonia fertilizers increase soil pH which in turn affects diversity, but in general, synthetic fertilizers reduce diversity. Manure and compost based on manure tends to decrease soil pH while some organic amendments, such as biochar, can increase pH.

Microbe Myth: Fertilizer Kills Microbes

Fertilizer kills microbes! This myth is believed by many and proponents of organic gardening keep spreading it. It is not true.

The myth is based on two misconceptions: all synthetic chemicals are harmful, and fertilizer is the reason agricultural soil is degraded.

This idea that synthetic fertilizer is automatically harmful and natural/organic is perfectly safe is completely unfounded. Caffeine is more toxic than the insecticide imidacloprid (a neonicotinoid). Vinegar is more toxic than glyphosate, the active ingredient in Roundup. Ricin is one of the most deadly compounds on Earth and it is a natural compound from the castor bean. Arsenic is a natural chemical found in all soil and produce. You can't tell how toxic or dangerous a chemical is by how it is made.

Agricultural soil is less fertile than natural soil in a forest or a meadow, but that does not mean synthetic fertilizer causes the change. Tilling, monoculture, removing produce, and pesticides all contribute to the change.

The reality is that microbes use nitrates, phosphates, and potassium just like plants. Fertilizer is microbe food. When added to soil in reasonable amounts, microbe populations increase.

If fertilizer killed microbes, they would also be killed by manure and compost because they provide the exact same nutrients as fertilizer.

The addition of excess nitrogen or phosphate, either synthetic or organic, can affect population dynamics, and several of these situations are discussed in this book. But fertilizer does not kill microbes.

Both synthetic and organic fertilizer affects the microbial community composition, diversity, and relative abundance, and as gardeners, we have almost no knowledge of how such changes affect our plants. It is therefore difficult for gardeners to make specific claims for any one type of treatment. The chemical nature of the amendment, the type of soil, and the environment all have an impact and gardeners have no reliable way to monitor changes in the microbe community.

In general, both forms of fertilizer are beneficial to the microbe biomass and both will alter the community. Both also lead to improved plant growth. Excessive amounts of either cause environmental issues, and excessive use of nitrogen reduces both bacterial and fungal biomass as well as the F:B ratio. Organic fertilizer has the added benefit of improving soil structure and increasing carbon (organic matter) levels in soil. Synthetic fertilizer has a more negative impact on the environment.

Crop Rotation

Crop rotation is a process where a given field is used for one crop followed by a different crop. In many systems each crop is grown in a different year, but in areas with longer growing seasons two crops could be grown in the same year.

Cover cropping is a special form of crop rotation where one crop is grown to partial maturity before it is incorporated into the soil. This adds green organic matter into the soil.

Plants cultivate and affect the soil microbes, so it makes sense that crop rotation can be used to develop the right microbe population for the next crop in the cycle. This has been practiced in a rudimentary fashion by farmers for many years, without knowing why certain crops helped control disease in the following crop.

Microbial diversity usually increases with crop rotation, but the increase is not nearly as large as you might guess, with values reported in the 5 percent range. Crop rotation only increases diversity when the right combinations are used. For example, rice/mung bean and maize/wheat rotations produced increased microbial richness, while a wheat/field pea rotation showed a decrease.

Crop rotation can both attract and repulse plant pathogens. Short rotation cycles tend to have a limited benefit on pathogens, while long cycles provide enough time for pathogen densities to decrease. This is even more effective if nonhost crops are inserted in the long cycles. In either case, crop rotation can be beneficial for controlling disease but it does not totally eliminate soil-borne pathogens.

Soil nutrient levels can also be positively affected by crop rotation. Adding one or more crops in a rotation can increase soil

Microbe Myth: Crop Rotation Works in Gardens

Crop rotation in a vegetable garden means not planting the same crop or member of a plant family in the same location every year. Some people use a three-year rotation where the same crop is not grown in the same spot until the fourth year. Others use a four-year rotation, or even a seven-year rotation. Some will leave the ground fallow in one of those years, and others will plant a cover crop for the full year.

Does crop rotation work? It definitely works in agriculture, but remember that farmers are rotating crops across a long distance. Tomatoes grown in one spot this year will be hundreds of acres away next year. This can confuse pests.

A gardener, on the other hand, moves the tomatoes ten to twenty feet away the second year. That won't confuse most pests. Some pests such as corn earworms, cabbage loopers, and potato leafhoppers travel hundreds of miles to a winter home and then return in spring. Crop rotation does not control them, even in agriculture.

The other reason for crop rotation is to change the nutrients plants extract from soil. Each plant uses a slightly different nutrient profile. This makes sense for agriculture, but gardeners tend to overfeed and they don't test their soil, making crop rotation less valuable.

One of the main reasons for crop rotation is to reduce diseases, especially ones where the microbe overwinters in soil. When these organisms become active in spring, they need to find a host plant. If none exists nearby, they tend to die out. In three to five years when the crop returns to the spot, the pathogen level will be much reduced. This is much less effective in gardens due to short distances.

Except in very specific cases, crop rotation adds very little value in a backyard garden.

carbon by 4 percent and total nitrogen by 5 percent. These numbers increase even more to 9 percent and 13 percent respectively when a cover crop is used.[22]

Plant Breeding

Current crops have been bred in soil that is fumigated and supplied with the luxury of high nutrient levels. Such breeding almost certainly selects for plants that are less efficient at promoting and supporting symbionts. There is evidence that newer crops are less able to grow in environments that are not as ideal as the breeding ground, but this has yet to be proven.

For example, breeding new varieties of lettuce has resulted in reduced branching in roots, making it more difficult for them to find nutrients. Cultured lettuce does not need large roots because agriculture brings nutrients to the roots. Over one hundred years of breeding has produced barley plants that release 32 to 85 percent less allelopathic exudates, making them less competitive with weeds. Newer cultivars of rice produce different exudates than older varieties, affecting the root population of ammonia-oxidizing bacteria (AOB). Since rice grows under flooded conditions, they rely on AOB to provide nitrogen.

Modern plant breeding has produced plants that rely less on microbes and more on chemical fertilizers. As we learn more about plant-microbe interactions, we should be able to reverse this trend and breed plants that grow much better with lower levels of fertilizer or pesticides. We can use the natural warring tendencies of microbes to benefit agriculture.

Biodiversity

Biodiversity of both the plants and the microbes are important.

The effect of plant biodiversity can be measured by using test plots of one, two, four, eight, or sixteen different plant species and examining various parameters. Increasing the number of species in a plot results in the following:

- increase in the microbial biomass
- soil improvement to deeper levels
- improved drought tolerance
- reduced effects due to waterlogging

Non-drought tolerant plants growing next to drought tolerant ones were better able to survive dry periods. A diverse mixture of plants results in a more diverse microbe community because each plant cultivates different microbes. As a group, both the microbes and the plants are better able to survive environmental extremes including heat, cold, and drought. A higher microbe diversity also leads to better soil aggregation, higher nitrogen fixation, and increased disease resistance.

Plants function best when there are many different kinds of microbes around their roots, and that happens when many different plant families grow together. Root mingling is good for the whole community.

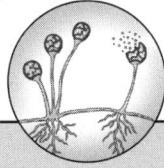

CHAPTER 13

Bioinoculants for the Garden

SELLING MICROBES IN A BOTTLE has become big business. Various terms are used for these products including bioinoculant, biofertilizer, biostimulant, and probiotics. The most popular term right now is inoculant.

In general, these are microbes that enhance plant growth by being applied to soil, to seeds before planting, or as a foliar spray. The vast majority of sales are for nitrogen-fixing products with phosphate solubilizing products a distant second. All of the others make up a small part of the 250 million dollar USA market. Most of the products sold today contain fungi and/or bacteria because these are more easily dried and packaged.

It is quite clear that microbes help plants grow better, so it is only natural for companies to sell them. There are now many inoculants for gardens, and I expect the number to rise significantly in the next few years. One problem with these products is that in most cases, it is impossible for the gardener to verify that the product has worked. In part this is due to the fact that claims are too general and really can't be tested, and secondly, the consumer has no way to see or identify microbes.

This chapter looks at the efficacy of some of the more popular classes of inoculants as well as a popular DIY inoculant, compost tea.

Bioinoculants for Seeds

Seed biopriming is a term used to describe the process of coating seeds in microbes before they are planted. The use of nitrogen-fixing rhizobium bacteria on legume seed is a widely accepted method. The addition of other bacteria at the same time is becoming more popular.

Adding rhizobium inoculants to legumes does work and has been shown to increase nitrogen fixation from 30 to 70 percent in grain legumes. However, strains of indigenous rhizobia are usually more efficient at colonizing nodules than inoculated strains, and they resist invasion by the commercial organism.

It is also important to remember that the rhizobium species has to be matched to the legume. For example, peas and beans are associated with different species. In order for legumes to form nodules and host the bacteria, the bacteria must be present in the soil. If your soil does not contain the right strain, nodules will not be formed.

Gardeners solve this problem by inoculating seed with the right bacteria at the time of planting. Little packs of bacteria can be purchased from seed companies or you can buy seed that is already coated with the right bacteria.

The product does not have to be applied every year, but it is a good idea to use it in a new garden. Once the bacteria is in the soil, it will survive for several years, so even a four-year crop rotation does not need to be inoculated each time.

How do you know if you have the right bacteria in the soil? Grow the legume and have a look at the roots halfway through the summer or in early fall. You can easily see the pea-size nodules if they are there. They are largest when the plant blooms. If the plant didn't make nodules, you either do not have the right bacteria in the soil, or you have too much nitrogen. Excess fertilizer will prevent the formation of nodules since the plant simply does not need the bacteria.

Seeds can also be inoculated to prevent specific diseases, and there is great interest in this as it would dramatically reduce the use of pesticides. For example, applying *Pseudomonas fluorescens* to seed suppresses infection on tomatoes by *Pythium ultimum*.

Seed inoculants for disease are not yet a viable option for gardeners for several reasons. It requires the correct identification of the disease, and gardeners don't normally get their plants tested.

A gardener also reacts to diseases on the current year's crop and by then it is too late to treat the seed. It is an option for persistent diseases that appear every year, but it is mostly used in agriculture.

Bioinoculants as a Foliar Spray

Several commercial foliar sprays are available for controlling diseases on tomato, celery, bell pepper, and citrus. In most cases, a single organism is applied to control an identified disease.

Applying inoculants as a foliar spray to control diseases has shown some positive results, but they are rare. In most cases, it is better to apply the product at the seedling stage so the microbes can grow with the plant.

Bioinoculants for Soil

Most of the inoculants available to gardeners are for use on soil. The sales pitch goes something like this:

"Microbes are important for soil health because they help turn organic matter into plant-available nutrients which produce better growing plants. Add this product to your soil and you will have healthier soil and plants."

Did you catch the marketing trick in this sales pitch? No? Read it again and see if you can catch it.

The first sentence is true and science has confirmed this many times. The second statement seems obvious given the first sentence, but it is rarely true.

You are probably asking yourself, if microbes are good for soil, why would adding them not be good for soil?

There are many reasons why adding commercial microbes might not be good for soil. Here are some of them:

- the microbes in the bottle are no longer living
- the microbes are a foreign species and should never be added to your soil
- microbes are always at capacity
- microbes are dynamic

Let's understand each of these, and then look at some specific cases.

Microbes Are No Longer Living

When you look inside a bottle of inoculants, all you see is some powder or a liquid, but you have no way to verify that the contents are as advertised.

The Oregon Department of Agriculture tested some products containing beneficial microbes. "Of the 51 products tested for bacteria, only nine met their guarantees. Of the 14 products tested for the fungi *Trichoderma*, **none** met their guarantees. Of the 17 products containing mycorrhizal fungi, only three met the guarantees made on the product label." DNA testing showed that in some cases the organism had never been in the product. In other cases, the organism was no longer viable.[23]

Remember that microbes are living organisms and they need to be cared for correctly. Take the case of mycorrhizal fungi, which is now a common inoculant. If the storage temperature gets too hot or too cold, the fungi die. If the nursery selling the product puts it in a hot greenhouse in summer, the fungi die.

Microbes Are Foreign Species

A lot of problems are caused by invasive species of plants and insects. These are relatively large organisms that we can easily see, even if we need to use a magnifying glass to see some of them.

What about microbes? They are too small to be seen by the naked eye, but there is no reason to think that non-native microbes might not become an invasive nightmare. Unfortunately, we still know so very little about soil microbes. We have not even identified 80 percent of them, and yet some people are quite confident that the microbes collected somewhere else and seeded in their soil will not cause a problem.

The reality is that we already know about pathogens that have caused problems after being moved into non-native areas. There is no reason to think that moving non-pathogens will be entirely safe. If you really want to add microbes to your soil, go to a local wooded area or local field and bring some of that soil home. A handful of that soil will add billions and billions of native microbes.

Microbes Are Always at Capacity

This is a concept that few gardeners understand and most marketing companies ignore. And yet it is critical for understanding microbe populations.

Microbes are always at capacity. What this means is that in any type of soil, the microbe population is always at maximum strength. Microbes populate an area so quickly that they are always filling any available space. Think of a football stadium where every seat is full. What happens if another one hundred people show up? They can't get in because all the seats are taken. The stadium is full to capacity.

Very poor infertile soil is like a small stadium with limited space and so it can only support a small number of people, but it is full. Really good fertile soil is like a large stadium that holds a lot of people, but even it is always full. Neither stadium can accommodate more people. Adding more microbes to soil is like sending people to a full stadium—they can't get in.

This is one of the fundamental flaws with soil inoculants. Promoters of the products try to convince you that your problem is a lack of microbes and that adding more will fix the problem. But the real problem is a lack of resources for the microbes you already have. Adding inoculants won't fix that.

Microbes Are Dynamic

Microbes reproduce very quickly when conditions improve, and when conditions are no longer acceptable, they die or go into a state of dormancy until things improve. Population densities change very rapidly.

The species that are present at any given time are also very dynamic. If soil gets wetter or warmer, new species start to thrive while others die off. Food sources have a great influence on populations. Most species can digest sugars, but only a few can survive on the lignin in woody plants. Oxygen levels and temperature are also important for species selection.

Science is just starting to get a handle on this, so as gardeners, we really can't tell what we have in our garden at any one time. We also have no way of knowing how to manipulate the populations to our advantage because we don't know when and if they change.

The fact that manufacturers of products claim to know exactly which microbes you should add to soil is quite surprising and should trigger all kinds of red flags. How can they know this when scientists don't?

Mycorrhizal Fungi

Mycorrhizal fungi inoculants have become very popular, and they are even added to most higher-end bags of potting mix and potting soil.

Most products contain between two and ten species. Scientists have identified 240 species that form associations with agricultural crops, not to mention those that work with ornamental plants. How do you know that the few species in a product are going to work with your plants? In most cases, you don't.

Consider a couple of examples from agriculture.

When AM is added to corn fields that have low compaction, the fungi improves plant growth. When the same product is added to corn growing in compacted soil, the fungi reduces plant growth, probably because it competes with the corn for resources.

Agricultural land in Argentina has a high level of native AM colonization in crop plants, even though the land has been extensively cultivated for many years. There is no value in adding an inoculant because the land has a high diversity of species and a high inoculum potential.

I have spoken with three different mycologists (scientists who study fungi), and all three agree that there is limited value in adding commercial mycorrhizal fungi to either containers or garden soil.

Should Gardeners Use Commercial Bioinoculants?

There is almost no scientific evidence that adding commercial microbes to soil will increase your microbe population. Microbes are always at capacity. The best way to increase the number of

Microbe Myth:
If It Is Good for Agriculture It Is Good for Gardens

A lot of the techniques and procedures used in gardening originate in agriculture. We gardeners like to mimic what the big boys do, thinking that if it works for them, it will work for us. The problem is that this leads to many poor practices in gardening.

Gardeners used to plant in well-separated rows just like an agricultural field until we learned we can plant much closer and get higher yields. Why? Gardeners don't use machinery that needs wide rows.

Agriculture uses crop rotation where they move crops miles from where they grew last year. Gardeners tried to copy this, and moved their tomatoes twenty feet from where they grew last year, and found crop rotation does not work in small gardens.

Foliar feeding is a common practice in agriculture, and it works well if done correctly and for the right reason. It does not work as a general fertilizing method in the garden or to solve unidentified disease issues.

Farmers do a lot of testing. They get soil tests done and then apply the right amount of fertilizer. Gardeners want to use fertilizer too, but they usually just buy a bag of 10-10-10 and spread it without knowing if they even need the nutrients.

The same issue applies to most inoculant products. Agriculture grows specific crops, have specific identified diseases, and use a product that was tested for their climate and specific soil type. Gardeners hear about this and apply it in a general way for many crops and many diseases in the hopes it will work. It rarely does.

The reality is that there are some interesting inoculants that do work in very specific conditions. They require proper testing to match the inoculant to the problem. Gardeners are not willing to do the testing, and for good reason. Who wants to spend hundreds of dollars to grow a few tomatoes? The bottom line is that most commercial inoculants should not be used by gardeners.

microbes is to increase the capacity—make your stadium bigger. You do that by adding more resources, and the best option for this is organic matter, which reduces compaction, increases moisture levels, and adds food. The microbes will quickly fill your new stadium.

Some foliar sprays do work for fighting disease, but in most cases, the organism has to be matched to the crop and the disease. That is beyond the capability of most gardeners. Leave this for agriculture to sort out.

Inoculation of legumes with rhizobium bacteria does work and is easy to do for the gardener. Just make sure you use the right product for your crop.

Compost Tea

Compost tea has become very popular, and many gardeners think it is much better than just using compost. Compost tea is a DIY form of inoculant.

You might think you know what compost tea is, but it can be made in many different ways. It can be made from compost, manure, vermicompost (worm tea), and even weeds (weed tea). All of these are very similar from a chemical and biological point of view.

Some people brew tea to make a concentrated form of plant-available nutrients. The brewing process speeds up decomposition, releasing plant-available nutrients and concentrating them in liquid form. The liquid is then easily spread over a larger surface area.

A different camp of gardeners don't care about the nutrients. For them, it is all about the microbes. Brewing increases the microbe population and puts them into a liquid media—the tea. This can then be spread on the garden to increase the number of microbes in soil.

Tea can be made in two very different ways: aerobically and anaerobically. The term aerobic means that the tea is made in the presence of oxygen. You usually bubble air through the tea as it brews. When tea is made anaerobically, it is made without added oxygen. You simply let the smelly sludge sit in a pail. The method used for making the tea affects the type of microbes because microbes tend to favor one or other of these living conditions.

Are Compost Tea Microbes Beneficial?

One of the main reported benefits of compost tea is that the microbes are "beneficial." This can only be true if they are not pathogens. The problem is that unless you have a fairly sophisticated lab, you won't know this. Home gardeners have no way to know which microbes are in their tea, so they don't know if they are beneficial. Making tea aerobically does not guarantee that it is pathogen free, as is commonly believed.

Does Compost Tea Add Nutrients?

Organic matter contains nutrients, but most of them are tied up in larger molecules. As the organic matter decomposes, the nutrients are released in plant-available forms. The starting material does contain nutrients, and brewing the material probably speeds up the release of nutrients.

The important question that needs to be asked is, does brewing tea produce more nutrients than in the original compost or manure? The answer to that is a very clear no. The brewing process cannot increase the total amount of nutrients. It might speed up their release from organic matter, but I suspect the effect is small.

Compost tea does not make the compost "go further," as so many claim.

Does Compost Tea Reduce Plant Diseases?

There has been some interesting science done on this that does show a reduction of some soil-borne diseases when the compost tea contains the right kind of microbes and it is used in the right climate and the right soil.

The scientists working on this test for specific microbes, and there is some future hope that this will result in inoculant products for specific diseases, but this work is far from conclusive and is not something a gardener can tackle.

The general claim that homemade compost tea reduces plant diseases is false.

Does Compost Tea Add Microbes to Soil?

There is no doubt this is true. If you have a pail full of slimy microbes and spread it around the garden, you are certainly adding microbes to the garden, but how long do they live? Remember, microbes are always at capacity, which means the newly added microbes are outcompeted and will die.

Can Compost Tea Be Dangerous?

When you make tea, you are creating an incubator for microbes by providing moisture, food, and the right oxygen level. But which microbes are you growing? You really have no idea.

The reality is that along with the "good" microbes you might also be growing "harmful" ones that can make you or your plants sick. Aerated tea can contain *Salmonella* and *E. coli*, both of which can be deadly to humans. Adding molasses to kick-start the tea actually increases the chance of brewing these pathogens. You could also be growing microbes that are harmful to plants.[24]

The process for making compost tea is not selective—you grow whatever is in the pot. I am confident that the risk for growing pathogens is low, but why take the risk when the benefits of compost tea are, at best, minimal?

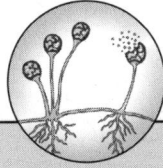

CHAPTER 14

Pathogens

MOST OF THIS BOOK has been about all of the good things microbes do for plants, but most gardeners are more concerned about the pathogens. When things are growing well, nobody thinks about the microbes, but the minute something doesn't look right, the search begins for either an insect problem or a pathogen disease.

The good news is that serious diseases in the garden are rare. I have been gardening for a long time and have grown a lot of plants. I have only encountered a few serious diseases. The ones that come to mind include:

- late blight in tomatoes robbed my whole crop for a couple of years
- one of my fifteen magnolia trees got verticillium wilt and died
- damping off disease kills seedlings quickly, but was easily prevented with proper cultural techniques and cinnamon
- found a virus on two hostas—both were discarded quickly
- rhizosphaera needle cast is affecting many of my spruce trees

Various fungal and bacterial rots have probably killed a number of bulbs and alpine plants without me ever identifying the problem. I also get some common problems such as powdery mildew on lilacs and black spot on roses, but I don't consider these serious since they don't kill the plants.

Many of the spots and deformed leaves you see are caused by abiotic (cultural) conditions, not a pathogen.

I have some good advice for beginning gardeners. Don't read too much about diseases. It will scare you into never gardening. You will never see most of the diseases that exist.

You Suspect a Disease—What Now?

One of the biggest mistakes people make when they suspect a problem is to get online and ask for a cure. The question usually goes something like this: "What can I spray to save my plant?"

You will get lots of answers, but none will be right because you have not yet identified the problem. Unfortunately, many gardeners want to be helpful and give out solutions before they understand the problem. Instead, follow these four steps for both pests and diseases.

Step 1: Identify the problem. Until you know what the problem is, you can't possibly find a solution. If you can't see the problem, you can't identify it.

Step 2: Research the problem. Understand the issue. Learn about the life cycle of the pest. How does it live, how does it reproduce, what does it eat, and most importantly, how big of a problem is it? Does it have natural predators that will take care of the problem for you?

Step 3: Decide if the problem needs a solution. Most problems in the garden do not have to be solved by the gardener. Many are not serious enough to bother with, and many resolve themselves.

Step 4: Find a solution that really works. Most DIY solutions recommended online do NOT work.

List of Plant Diseases

The following table is a list of some of the more common diseases.

Name (common/ botanical)	Type	Description	Control
Alternaria leaf spot (*Alternaria* spp.)	Fungus	Fairly large brown spots on leaves that turn black as the fungus produces spores. Lower leaves are affected first. Develops mostly in summer and can cause leaf drop. The disease develops when dew forms, humidity is high, and air is stagnant.	Controlled by fungicides and cultural means such as good air circulation.
Anthracnose (various species) Also called leaf, shoot, or twig blight	Fungus	Anthracnose describes a group of related host-specific fungal diseases that typically cause dark lesions on leaves. Also causes sunken lesions and cankers on twigs and stems of both deciduous and evergreen woody plants. Also infects vegetables, flowers, fruit, and turfgrass.	Fungicides have limited value for prevention, and once symptoms develop it can't be effectively controlled. Pruning infected parts may eliminate the disease. The fungus can remain in soil for many years after removal of the plant.
Armillaria root rot (*Armillaria mellea*)	Fungus	Trees with the disease have smaller chlorotic leaves that drop early. Upper canopy is thin because shoots die back. Plants grow slowly. Mycelium may be visible under the outer bark.	Infected trees are best removed. Deeply planted trees or trees with soil covering the root collar area are more susceptible. There is no known cure, but native fungi in soil do control the disease. Don't over water.
Bacterial canker (*Pseudomonas syringae*)	Bacteria	A disease that affects the genus *Prunus* (cherry, plum, and peach). Branches die back, infected leaves become spotted and yellowed, cankers produce a gummy resinous ooze, and wood in the cankered area is discolored. Entire tree can die.	Keep trees healthy and prune out any affected wood. Copper-containing sprays may help, but copper-resistant *Pseudomonas* are becoming more common.

Name (common/ botanical)	Type	Description	Control
Bacterial soft rot (*Erwinia* spp., *Pseudomonas* spp., *Bacillus* spp., and *Clostridium* spp.)	Bacteria	A general term for many diseases affecting a wide range of plants. Plant cell structure is degraded by the bacteria to form water-soaked spots that eventually enlarge and become soft and sunken. Underlying tissue becomes mushy with a strong rotting odor.	Once plant tissue is infected, there is no cure. Cut out and discard the infected part of the plant. Avoid wet conditions, don't crowd plants, and prevent damage to tissue so the pathogen can't gain entry.
Botrytis rot, gray mold (*Botrytis cinerea*)	Fungus	Also called Botrytis fruit rot since it affects mostly fruit either as it ripens or even in the flowering stage. Grows well as fruit develops higher sugar content. Starts as small lesions that sporulate in a couple of days, spreading quickly. Looks like gray fuzzy mold.	Very prevalent and always present. Removing infected leaves and fruit reduces spore level. Survives in soil and grows best in cool, wet conditions. Can be managed with hygienic cultural controls and cultivar selection.
Brown rot (*Monilinia fructicola* and *M. laxa*)	Fungus	Disease of *Prunus* trees that affects mostly the fruit. In spring, brown spots develop on blossoms that fail to produce fruit. Fruit that does develop forms a brown powdery fungal growth that engulfs the fruit, which shrivels up.	Brown rot is not fatal, but infected fruits can't be saved. Prune out infected wood. Remove all infected material. Fungicides can be used for chronic problems.
Clubroot (*Plasmodiophora brassicae*)	Fungus	Disease affects the cabbage family. Plants are stunted, wilt easily, and have yellowing leaves. Roots are swollen into thick, irregular club shapes. Soil-borne.	Prevention is the best option. Use disease-free transplants or seeds. Fungicides are of limited value.
Crown gall (*Agrobacterium* spp.)	Bacteria	Causes rough, woody, tumor-like galls to form on roots, trunks, and branches of woody plants. Reduces plant growth and vigor. Can kill young plants, but large trees can tolerate the disease.	Prevention is best. Inspect all new plants. Drench potted plants with *Agrobacterium radiobacter* K-84, a biological control that produces an antibiotic.

Name (common/ botanical)	Type	Description	Control
Damping off (*Rhizoctonia* spp., *Fusarium* spp., *Pythium* spp.)	Fungus	A disease of seedlings. Thrives in cool, wet conditions. Spreads quickly from plant to plant. Seedlings suddenly fall over and shrivel up.	Drier soil with a fan running 24/7 prevents the disease in most cases. Fungicide or cinnamon work to stop the spread.
Early blight (*Alternaria tomatophila*, *Alternaria solani*)	Fungus	Common on tomatoes and potatoes. Affects leaves, fruits, and stems. Dark spots form on older foliage. Round brown leaf spots with rings. Leaves go brown and fall off. Stems develop sunken spots. Usually does not affect an early harvest, but reduces yield.	Overwinters on foliage in soil. Disease spreads in spring from splashed soil. Prevent soil splash by covering soil with mulch. Remove lower leaves.
Fire blight (*Erwinia amylovora*)	Bacteria	Affects woody plants. Kills blossoms and shoots. Causes die back of branches. Can kill trees. Plants look as if they were scorched with fire. Leaves turn brown and black but remain attached to branches.	Prune out infected branches. Limit structural pruning, fertilize less. Pesticides for bacteria may help.
Fusarium wilt (*Fusarium oxysporum*)	Fungus	Affects a wide range of plants. Yellowing, stunting, and death of seedlings. Xylem tissues turn brown and the plant may die. Caused by host-specific forms of the same species.	Treat suspect seed with fungicide. Fusarium in the soil can be reduced by heat treatments and chemical fumigation.
Late blight (*Phytophthora infestans*)	Water mold	Devastating disease of tomato and potato, infecting leaves, stems, fruit, tubers. Spreads a long way by air. Late blight was responsible for the Irish potato famine of the late 1840s. Plants killed soon after infection in mid to late summer.	Started with infected tubers or transplants. Some resistant varieties exist. No cure once the plant is infected.
Leaf spot disease (various including *Septoria* spp.)	Fungus	Symptoms vary. Leaf spots have well-defined brown, black, or tan margins with reddish centers. Spots grow to cover the whole leaf. Overwinter on plant debris. Pathogens tend to be host specific and nonlethal.	Plant cleanup and healthy plants go a long way to control the disease. Fungicides are not usually warranted.

Name (common/ botanical)	Type	Description	Control
Mosaic viruses (various)	Virus	Term applies to any virus that causes infected plant foliage to have a mottled appearance— includes a couple of hundred species. Leaves are mottled yellow and white with light or dark green spots and streaks.	No cure; remove infected plants. Start with virus-free plants.
Pink snow mold, Microdochium patch or Fusarium patch (*Microdochium nivale*)	Fungus	Infects cool-season turfgrass under a cover of snow. Worse in years with a lot of snow. Pink, white, or tan patches of dead and matted leaf blades.	Keep turf mowed in late fall, use low nitrogen fertilizer in fall. Rake in spring, fertilize, and let regrow.
Powdery mildew (various species)	Fungus	Leaf surfaces are covered with a grayish white powdery growth. Starts as small spots enlarging to cover the whole leaf. Leaves may turn yellow in later stages of the disease. Mostly cosmetic. Rarely kills a plant.	More air flow may help. Keep plants well watered and healthy. Use resistant varieties.
Rust (various species)	Fungus	A variety of diseases with a common name. Leaves get covered with spores that are a reddish brown to orange. Easily rubs off by hand. Some rusts are mostly cosmetic, while others can do serious damage.	Keep plants healthy. Can be controlled with fungicides, but they are usually not needed. Some plant types get it every year, so it is better to replace them with other species or cultivars.
Verticillium wilt (*Verticillium* spp.)	Fungus	A serious, usually fatal disease of both woody and herbaceous. First signs are sudden wilting and die back. A cut stem shows brown or green streaking.	Soil-borne pathogen. No cure. Start with resistant plants.

Fighting Plant Diseases

Most diseases can only be controlled using synthetic pesticides. A few DIY solutions do work, and are listed below.

Carefully follow the labeled rates for any pesticide. Too many gardeners apply them at higher rates, thinking that this will work better, but it rarely does. Higher rates can actually be less effective.

DIY Pesticides

I am always amazed at the number of solutions that are recommended online, in blog posts, on YouTube videos, or in response to a question on social media. Ninety percent either don't work or have no scientific basis. And yet readers quickly accept any nonsense that is presented to them, especially if they already have the ingredients. Anything from the kitchen seems to be worth trying.

If someone recommends a home remedy but does not provide a link to a scientific study, it probably does not work. If the remedy is very general, like "garlic juice is a good fungicide," it almost never works. Remember that there are thousands of different fungi and no treatment works on all of them. If the solution includes Epsom salts, you definitely know it doesn't work.

The following is a short list of DIY solutions that are known to work on plant diseases.

Milk for Powdery Mildew

Milk does work, but there is a catch. It does not stop the disease or reduce the mycelium once it is growing on the leaves. Milk may slow down growth and it seems to prevent spores from germinating. Try using a 25 percent solution of 2% milk. Non-fat milk does not work as well. Contrary to so many claims, it has no effect on black spot disease on roses.

It also works on downy mildew on cucurbits. The key to using milk is to spray it on the leaves before you see any symptoms and then to spray weekly to replace the washed-off milk.

Baking Soda for Powdery Mildew

Baking soda will also work on powdery mildew, and this has been researched by numerous people including Cornell University, who developed the Cornell Formula for this. The exact ingredients are not published but it is similar to this:

- 1 gallon water
- 1 tablespoon baking soda
- 1 tablespoon vegetable oil
- 2 drops dishwashing liquid

These ingredients are mixed together and then sprayed on the plant that is under fungal attack. It is popular because it's simple and most people have easy access to the ingredients. If the oil is left out it still works but is less effective.

Baking soda is sodium bicarbonate, and sodium is toxic to plants. It is better to use potassium bicarbonate, but that is harder to source.

Chamomile Tea

I have used chamomile tea to stop damping off disease in seedlings and it seems to work. I have not found any scientific study that looked at this specific use of the tea, but the tea has been shown to have mild antibacterial and antifungal properties.

Cinnamon

You probably know cinnamon as a brown powder that you buy in grocery stores. It is made by grinding the bark of certain trees. What you probably don't know is that the stuff you buy as cinnamon is not the true cinnamon. The real stuff comes from a special species of tree and it's very expensive, so most companies use a cheaper imitation.[25] Various types of cinnamon have been tested and show antifungal properties.

Cinnamon can be used in a couple of ways. Sprinkling it on seedlings will prevent damping off disease, which is a fungal infection. Once seedlings are infected, it can also stop the progression of the disease to uninfected seedlings.

It can also be sprinkled on any cut surface to prevent infection. Orchids can get a bacterial disease known as crown rot. As soon as you see the rot, sprinkle a heavy dose on any parts showing infection. Make sure you get it right down into any crevices in the foliage. It stops the rot instantly.

Cinnamic acid is extracted from cinnamon and has been used in commercial antimicrobial products.

Neem Oil

Neem oil is pressed from the fruit and seeds of the neem tree (*Azadirachta indica*). It is effective at killing various rusts and powdery mildew spores, but it's less effective against rose black spot (*Diplocarpon roseae*) and other fungal diseases. It does kill virus vectors such as aphids and white fly.

The active ingredient in neem is azadirachtin. There are two types of neem. Pure cold-pressed neem preserves azadirachtin and is used as an insecticide. Neem is also available with the active ingredient removed and it is used in food, skin care, and cosmetic products. If you are buying neem, look for a product that specifies the azadirachtin concentration on the label.

Commercial Pesticides

Commercial solutions tend to work better than DIY solutions and have been tested for specific pests and pathogens.

The names for different types of pesticides are frequently misused. A pesticide is any product that kills or prevents any pest or disease including insects, fungi, bacteria, nematodes, etc. It is a general term, and each of the following are pesticides:

- fungicide—for fungi
- bactericide—for bacteria
- insecticide—for insects
- herbicide—for plants (weeds)
- rodenticide—for rodents

Products are available in both organic and synthetic formulations. In general, synthetic products are more effective and have been more widely tested for efficacy and safety. Organic products have a more general use but are less effective. Contrary to popular belief, many have had very little testing.

There are two classes of pesticides: preventatives and curatives. Preventative products are applied before the pest arrives and the treatment prevents the pest from getting started. Many of these are systemic. Curatives are applied after the pest is present. They generally have little effect unless they can make direct contact with the pest.

Fungicides

Fungicides are pesticides that prevent, kill, or inhibit the growth of fungi on plants. They are not effective against bacteria, nematodes, or viruses.

Contact fungicides, also called protectants, are not absorbed by the plant, but they do stick to the surface of the plant for a limited time period. They provide a protective layer that prevents the fungi from attacking the plant. These products are effective but need to be reapplied as they wear off.

Systemic fungicides are absorbed by the plant and once inside are transported to all parts of the plant. Some products are even transported into new growths to provide long-term protection.

The timing of application is important. Preventive fungicides need to be applied before there is visible growth of the fungal disease. Many of these are applied in spring as new plant tissue is formed and then reapplied periodically. Curative fungicides can be applied after the disease starts since they are able to kill the fungi after it infects the plant.

Fungicides can work in a number of different ways. They can damage the fungal cell membranes or inhibit a critical chemical process. No fungicide is effective against all fungi.

Fungi can develop a resistance to chemical treatments, so it is a good idea to alternate between products to help prevent the development of resistant strains.

Organic Fungicides

The following are some organic fungicides that are available as commercial products.

Sulfur

This fungicide has been used for over two thousand years and it is still a viable option. It is effective against powdery mildew, rose black spot, rusts, and other diseases. Sulfur prevents spores from germinating, so it must be applied before the disease starts.

Sulfur is available as a dust, wettable powder, or liquid form. It should not be applied with or soon after applying an oil spray since the combination can harm plants.

A lime-sulfur spray is a mixture of lime (calcium hydroxide) and sulfur and is a common dormant spray that is applied before bud break in spring. It is more effective than sulfur alone.

Copper

Copper kills fungi and bacteria, but it can also harm plants. It has been used for a long time in a formulation called a Bordeaux mixture, which combines copper sulfate and lime. It is a popular spring spray because it adheres well even in rains. This product is effective for a wide range of diseases but always follow label instructions closely to minimize plant damage.

Bactericides

Bacteria can cause a number of serious diseases in plants, such as fire blight on pear and apple, and bacterial spot on peach, but there are limited options for fighting them. The above-mentioned Bordeaux mixture is one option. The other option is the use of antibiotics, but these are generally not available to gardeners.

Available antibiotics include things like streptomycin and oxytetracycline. Both products are more highly regulated than pesticides.

The best option for gardeners is to grow healthy plants.

Human Diseases

Gardening is considered to be a very healthy exercise, but it can be a source of human diseases. The two main areas of concern are soil and produce.

Soil-borne Diseases

You don't hear about these diseases very often, but there are quite a few diseases you can get from soil, compost, and even peat moss.

They can come from garden soil and even so-called sterile bagged soil. Here is a list of some of the more common ones:[26]

- **Valley Fever** occurs when people inhale fungi that belong to the group Coccidiodes, which are found in the southwestern United States. The tiny spores live in desert dirt, and on windy days they can get blown around and inhaled. Severe cases can lead to pneumonia.
- **Hantavirus** has a high mortality rate and is spread by rodent droppings, urine, and saliva. It can become airborne and infect gardeners.
- **Tetanus** causes about twenty-five deaths in the USA every year and many more in Asia, Africa, and South America. The bacteria causing tetanus is common in soil, dust, and feces.
- **Botulism** is well-known as a disease related to infected food, but "wound botulism" can be contracted from soil. It is a bacterial infection.
- **Brain-eating amoeba** kills almost everyone that is infected. The single-celled microbe is found in warm freshwater and needs to enter the body through the nose. Unless the gardener goes for a swim in their pond, they should be safe.
- Several strains of ***Escherichia coli*** (*E. coli*) cause diseases. One strain, ETEC, accounts for several hundred million cases of diarrhea and tens of thousands of deaths globally each year.
- **Melioidosis** is a bacterial infection that quietly causes thousands of deaths each year as people come in contact with mud.
- **Histoplasma** is a fungus that causes lung infections and is spreading throughout the US and probably other countries as well.

There are lots of diseases and lots of potential exposures, but to really understand the risk we have to look at rates of infection. The European Union shows about thirty-one infections (includes all soil-borne diseases), per one hundred thousand people per year. It is possible that the rate is a bit higher in gardeners because they are exposed more. To put this into perspective, five thousand out of one hundred thousand get influenza each year, resulting in eight deaths, and we don't all go around wearing gloves and masks to prevent catching it.

The above-mentioned organisms are found in soil, but they need to enter your body to cause infection. This can happen through a cut in the skin, through the mouth, or through the nose. Many air-borne fungal and bacteria organisms travel easily through the air. If you don't touch the soil and you stop breathing, you are safe!

The risk of these diseases is low and should not be a concern. To reduce the risk of infection even more, wear gloves, keep cuts covered, wear closed shoes, and wash often; but to be honest, there is very little science to confirm the effectiveness of these measures.

Diseases from Eating Produce

You can get sick from eating commercial produce. What many people don't realize is that you can also get sick from eating your own homegrown produce. Any food that comes in contact with soil or the air above soil can contain pathogens.

The best way to reduce your risk of sickness is to wash all produce in running water. It works as well or better than all of the other options.

Endnotes

Chapter 1: Introduction

1. Brendan B. Larsen, Elizabeth C. Miller, Mathew K. Rhodes, John J. Wiens, "The Number of Species on Earth and the New Pie of Life," *The Quarterly Review of Biology*, 92, no. 3, (September 2017).

Chapter 2: The World Under a Microscope

2. J. Hoorman, R. Islam, "Understanding Soil Microbes and Nutrient Recycling," *Environmental Science*, (2010).
3. Robert Pavlis, "Sterile Soil—Does It Really Exist?," *Garden Myths* (blog).

Chapter 3: Bacteria

4. Akshit Puri, Kiran Preet Padda, and Chris P. Chanway, "Nitrogen-Fixation by Endophytic Bacteria in Agricultural Crops: Recent Advances," *Nitrogen in Agriculture*, (2018).

Chapter 4: Fungi

5. Images from various sources:
en.wikipedia.org/wlki/File:White_fungus_in_wood_chips.jpg
commons.m.wikimedia.org/wiki/File:Dog_Vomit_Slime_Mold_(Fuligo_septica)_-_Guelph,_Ontario_2020-06-19.jpg
6. Images from various sources:
commons.wikimedia.org/wiki/File:Slice_of_bread_with_mould_%281%29.jpg
commons.wikimedia.org/wiki/File:Rhizopus_stolonifer4.JPG
www.flickr.com/photos/gjshepherd/3226473328
7. Images from various sources:

www.rawpixel.com/image/3322948/free-photo-image-agaric-amanita-berkeley

commons.wikimedia.org/wiki/File:Morchella_esculenta_41997.jpg

www.flickr.com/photos/fungi-nis/5830470879

www.flickr.com/photos/ivanteage/4398993327

8. Aline Fernandes Figueiredo, Jens Boy, and Georg Guggenberger, "Common Mycorrhizae Network: A Review of the Theories and Mechanisms Behind Underground Interactions," *Frontiers in Fungal Biology*, (September 2021).

Chapter 5: Yeast

9. Elena Sláviková, Renáta Vadkertiová, Dana Vránová, "Yeasts Colonizing the Leaves of Fruit Trees," *Annals of Microbiology*, 59(3): (September 2009): 419–424.

Chapter 6: Nematodes

10. Robert Pavlis, "Will Marigolds Stop Root Knot Nematodes?," *Garden Myths* (blog).

11. Robert Pavlis, "Garlic Nematodes (Bloat)—A Real Threat to Your Garlic Crop," Garden Fundamentals, October 10, 2018, video.

Chapter 8: Viruses

12. M. D. García-Pedrajas, M. C. Cañizares, J. L. Sarmiento-Villamil, A. G. Jacquat, and J. S. Dambolena, "Mycoviruses in Biological Control: From Basic Research to Field Implementation," *Phytopathology*, 109 (2019): 1828–1839.

Chapter 9: More Microbes

13. Robert Pavlis, "Is Soil an Antidepressant—Does It Make You Feel Good?," *Garden Myths* (blog).

Chapter 10: Microbe Communities

14. Images from various sources:

www.flickr.com/photos/emsl/15544641432

www.flickr.com/photos/microbeworld/5981923914
www.flickr.com/photos/pnnl/23895786255
www.flickr.com/photos/pnnl/8146290867

Chapter 11: Plants Love Microbes

15. Government of Canada, "Produce Safety," February 28, 2023.
16. "Best Way to Wash Fruits and Vegetables," *Garden Myths* (blog).
17. James F. White, Kathryn L. Kingsley, Qiuwei Zhang, Rajan Verma, Nkolika Obi, Sofia Dvinskikh, Matthew T. Elmore, et al., "Review: Endophytic Microbes and their Potential Applications in Crop Management," *Wiley Online Library* (July 27, 2019).
18. Robert Pavlis, "How to Save Tomato Seeds," Garden Fundamentals, January 19, 2019, video.
19. Eman M. Khalafi, Manish N. Raizada, "Bacterial Seed Endophytes of Domesticated Cucurbits Antagonize Fungal and Oomycete Pathogens Including Powdery Mildew," *Frontiers in Microbiology*, (February 5, 2018).

Chapter 12: Manipulating Microbes

20. Xiaoli Wanga, Weixin Zhangb, Yuanhu Shaob, Jie Zhaoc, Lixia Zhoud, Xiaoming Zoue, and Shenglei Fu, "Fungi to Bacteria Ratio: Historical Misinterpretations and Potential Implications," *Acta Oecologica*, 95 (2019): 1–11.
21. A. Muller, C. Schader, N. El-Hage Scialabba, et al., "Strategies for Feeding the World More Sustainably with Organic Agriculture," *Nat Commun* 8, (2017): 1290.
22. M. D. McDaniel, L. K. Tiemann, A. S. Grandy, "Does Agricultural Crop Diversity Enhance Soil Microbial Biomass and Organic Matter Dynamics? A Meta-analysis," *Ecological Applications*, 24, no. 3 (April 2014): 560–570.

Chapter 13: Bioinoculants for the Garden

23. Alexis, "ODA Finds Big Problems with Little Organisms," *Oregon Department of Agriculture*, (2016).
24. Brion Duffy, Chester Sarreal, Subbarao Ravva, Larry Stanker, "Effect of Molasses on Regrowth of *E. coli* O157:H7 and

Salmonella in Compost Teas," *Compost Science & Utilization,* 12
(1), (2004).

Chapter 14: Pathogens

25. Robert Pavlis, "Cinnamon—Does It Stop Damping Off in Seedlings?," *Garden Myths* (blog).
26. Jeffery Simon Lee, Van der Putten, "Soil Borne Human Diseases.
EUR 24893," *Publications Office of the European Union,* (2011).

Index

About the Author

ROBERT PAVLIS, a Master Gardener with over forty-five years of gardening experience, is the owner and developer of Aspen Grove Gardens, a six-acre botanical garden featuring three thousand varieties of plants. A popular and well-respected speaker and teacher, Robert has published articles in *Mother Earth News*, *Ontario Gardening* magazine, a monthly plant of the month column for the Ontario Rock Garden Society website, and local newspapers. He is also the author of two widely read blogs: GardenMyths.com, which explodes common gardening myths, and GardenFundamentals.com, which provides gardening and garden design information. Robert also has a gardening YouTube channel called Garden Fundamentals.

Connect with Robert Pavlis

You can connect with me through social media by leaving comments at one of the following:

http://www.gardenmyths.com
http://www.gardenfundamentals.com
http://www.youtube.com/Gardenfundamentals1

The best way to reach me directly is through my Facebook Group: https://www.facebook.com/groups/GardenFundamentals, where I answer questions on a daily basis.

Also by the Author

Compost Science for Gardeners

With breathtaking clarity, *Compost Science for Gardeners* demystifies composting practices and helps readers determine the best technique for their unique situation. This comprehensive science-based book is your key to building healthier soil and growing better plants.

Using plain language and easy-to-follow instructions, this essential resource distills and blends the latest scientific research with the author's many decades of knowledge and experience into manageable form, debunking a host of common gardening myths along the way. Learn about:

- the role of composting in the ecological cycle
- compostable materials to incorporate and those to avoid
- browns and greens VS the carbon-to-nitrogen ratio
- the full range of composting methods, including cold and hot composting; composting in piles, bins, and tumblers; and pit and trench composting
- keyhole gardening, food digesters, vermicomposting, bokashi, eco-enzyme fermentation, and more
- the relative merits and impact on the environment of each composting technique
- using finished compost to improve soil health.

Anyone can compost. Whether you are a balcony or backyard gardener, market gardener, small-scale farmer, or homesteader, or even if you are simply looking for a way to keep organic matter out of the landfill, this book will show you how to do it simply, safely, and sensibly.

Plant Science for Gardeners

Plant Science for Gardeners empowers growers to analyze common problems, find solutions, and make better decisions in the garden for optimal plant health and productivity.

By understanding the basic biology of how plants grow, you can become a thinking gardener with the confidence to problem-solve for optimized plant health and productivity. Learn the science and ditch the rules! Coverage includes:

- the biology of roots, stems, leaves, and flowers
- understanding how plants function as whole organisms
- the role of nutrients and inputs
- vegetables, flowers, grasses, and trees and shrubs
- propagation and genetics
- sidebars that explode common gardening myths
- tips for evaluating plant problems and finding solutions.

Whether you're a home gardener, microfarmer, market gardener, or homesteader, this entertaining and accessible guide flattens the learning curve and gives you the knowledge to succeed no matter where you live.

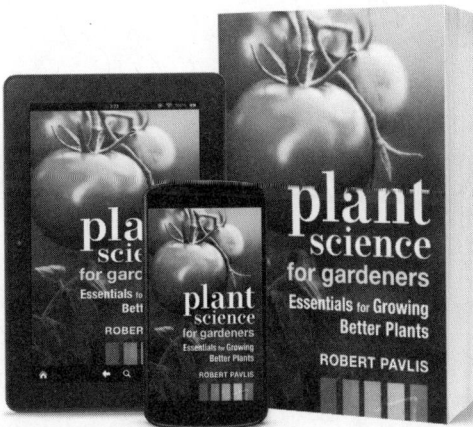

Soil Science for Gardeners

Robert Pavlis, a gardener for over four decades, debunks common soil myths, explores the rhizosphere, and provides a personalized soil fertility improvement program in this three-part popular science guidebook. Coverage includes:

- soil biology and chemistry and how plants and soil interact
- common soil health problems, including analyzing soil's fertility and plant nutrients
- the creation of a personalized plan for improving your soil fertility, including setting priorities and goals in a cost-effective, realistic time frame
- creating the optimal conditions for nature to do the heavy lifting of building soil fertility

Written for the home gardener, market gardener, and micro-farmer, *Soil Science for Gardeners* is packed with information to help you grow thriving plants.

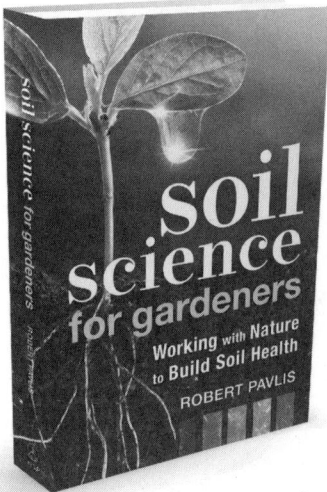

Garden Myths

If you enjoyed this book, you may also like Robert's other books, *Garden Myths Book 1* and *Garden Myths Book 2*. Each one examines over 120 horticultural urban legends.

Turning gardening wisdom on its head, Robert Pavlis dives deep into traditional garden advice and debunks the myths and misconceptions that abound. He asks critical questions and uses science-based information to understand plants and their environment. Armed with the truth, Robert then turns this knowledge into easy-to-follow advice. Details about the books can be found at http://www.gardenmyths.com/garden-myths-book-1/ .They are available from Amazon and other online outlets.

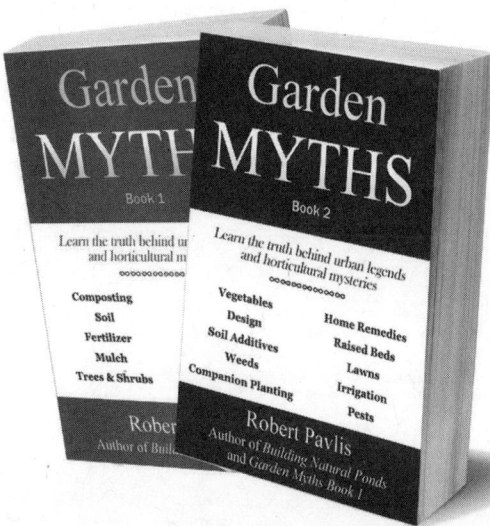

Building Natural Ponds

Building Natural Ponds is the first step-by-step guide to designing and building natural ponds that use no pumps, filters, chemicals, or electricity and mimic native ponds in both aesthetics and functionality. Highly illustrated with how-to drawings and photographs.

For more information and ordering details, visit:
www.BuildingNaturalPonds.com

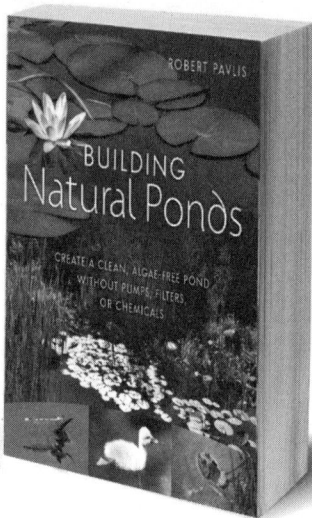

ABOUT NEW SOCIETY PUBLISHERS

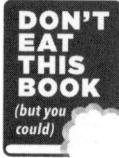

DON'T EAT THIS BOOK *(but you could)*

New Society Publishers is an activist, solutions-oriented publisher focused on publishing books to build a more just and sustainable future. Our books offer tips, tools, and insights from leading experts in a wide range of areas.

We're proud to hold to the highest environmental and social standards of any publisher in North America. When you buy New Society books, you are part of the solution!

+ This book is printed on **100% post-consumer recycled paper**, processed chlorine-free, with low-VOC vegetable-based inks (since 2002)
+ Our corporate structure is an innovative employee shareholder agreement, so we're one-third employee-owned (since 2015)
+ We've created a Statement of Ethics (2021). The intent of this Statement is to act as a framework to guide our actions and facilitate feedback for continuous improvement of our work
+ We're carbon-neutral (since 2006)
+ We're certified as a B Corporation (since 2016)
+ We're Signatories to the UN's Sustainable Development Goals (SDG) Publishers Compact (2020–2030, the Decade of Action)

At New Society Publishers, we care deeply about *what* we publish—but also about *how* we do business.

To download our full catalog, sign up for our quarterly newsletter, and learn more about New Society Publishers, please visit newsociety.com

ENVIRONMENTAL BENEFITS STATEMENT

New Society Publishers saved the following resources by printing the pages of this book on chlorine free paper made with 100% post-consumer waste.

TREES	WATER	ENERGY	SOLID WASTE	GREENHOUSE GASES
20	1,600	8	67	8,490
FULLY GROWN	GALLONS	MILLION BTUs	POUNDS	POUNDS

Environmental impact estimates were made using the Environmental Paper Network Paper Calculator 4.0. For more information visit www.papercalculator.org

FSC MIX Paper from responsible sources FSC® C016245 www.fsc.org

SDG PUBLISHERS COMPACT

Certified B Corporation

new society PUBLISHERS
www.newsociety.com